Name ______________________ Class ______________ Date ______________

Skills Worksheet

Directed Reading A

AF893764

Section: Sorting It All Out

Write the letter of the correct answer in the space provided.

______ **1.** What is the name for placing things into groups based on similar characteristics?
a. grouping
b. classification
c. studying
d. listing

WHY CLASSIFY?

______ **2.** What do scientists learn by classifying living things?
a. which living thing is strongest
b. the characteristics of different species
c. which species has the most members
d. how species get along

HOW DO SCIENTISTS CLASSIFY ORGANISMS?

______ **3.** Who founded modern taxonomy?
a. Albert Einstein
b. Charles Darwin
c. Isaac Newton
d. Carolus Linnaeus

______ **4.** What do taxonomists do?
a. only take photographs of animals
b. only name and photograph living things
c. describe, classify, and name living things
d. only take photographs of plants

Classification Today

______ **5.** How many levels of classification do most scientists use today?
a. eight
b. six
c. five
d. four

______ **6.** What do closely related living things share?
a. size only
b. shape only
c. many characteristics
d. color and size only

______ **7.** What characteristics do platypuses, brown bears, lions, and house cats all share?
a. giving birth to live young
b. retractable claws
c. ability to purr
d. hair and mammary glands

Branching Diagrams

______ **8.** What do house cats have that lions don't have?
a. ability to purr
b. retractable claws
c. hair and mammary glands
d. giving birth to live young

______ **9.** Which of the following pairs of animals are the most closely related?
a. lions and house cats
b. lions and platypuses
c. house cats and platypuses
d. house cats and brown bears

______ **10.** What characteristic is shared by bears, lions, and cats?
a. hooves
b. giving birth to live young
c. cold blooded
d. mane and tail

LEVELS OF CLASSIFICATION

Match the correct description with the correct term. Write the letter in the space provided.

______ **11.** The largest groups for classifying organisms are these.

______ **12.** Phyla are sorted into these.

______ **13.** Families are broken down into these.

______ **14.** Genera are sorted into these.

a. domains
b. species
c. classes
d. genera

SCIENTIFIC NAMES

Write the letter of the correct answer in the space provided.

______ **15.** What kind of name is specific to each living thing?
- **a.** common name
- **b.** nickname
- **c.** scientific name
- **d.** last name

Two-Part Names

______ **16.** What is the first part of a scientific name?
- **a.** the order name
- **b.** the genus name
- **c.** the kingdom name
- **d.** the species name

______ **17.** What is the second part of a scientific name?
- **a.** the order name
- **b.** the genus name
- **c.** the kingdom name
- **d.** the specific name

______ **18.** What is the scientific name for the Asian elephant?
- **a.** *Felis domesticus*
- **b.** *Elephas maximus*
- **c.** *Tyrannosaurus rex*
- **d.** *Canis lupus*

______ **19.** What do all genus names begin with?
- **a.** a lowercase letter
- **b.** a capital letter
- **c.** a number
- **d.** a Roman numeral

______ **20.** Scientific names are usually in what languages?
- **a.** Latin and French
- **b.** Greek and German
- **c.** Latin and Greek
- **d.** English and Latin

EXTINCT ORGANISMS AND LIVING ORGANISMS

______ **21.** How are the characteristics of extinct animals identified?
- **a.** from written histories
- **b.** from cave drawings
- **c.** these characteristics can't be identified
- **d.** from fossils of that organism

FOSSILS AND BRANCHING DIAGRAMS

______ **22.** When did Neohipparion appear?
- **a.** in the Pliocene
- **b.** in the Pleistocene
- **c.** in the Miocene
- **d.** in the Oligocene

Name ______________________ Class ____________ Date ____________

Skills Worksheet

Directed Reading A

Section: Domains and Kingdoms

Write the letter of the correct answer in the space provided.

______ **1.** What categories did people think all organisms fit into before organisms, such as euglena, were discovered?
a. plants or animals
b. fish or birds
c. plants or mammals
d. animals or trees

THREE DOMAINS

______ **2.** On what basis do scientists classify organisms?
a. geographically
b. alphabetically
c. unique characteristics
d. shared derived characteristics

______ **3.** What kingdom did scientists add for organisms that may have both plant and animal characteristics?
a. Euglena
b. Protista
c. Fungi
d. Animalia

______ **4.** Today, how many domains are in the classification system?
a. five
b. three
c. seven
d. eight

DOMAIN ARCHAEA

______ **5.** What is the name given to small, single-celled organisms with no nuclei?
a. seeds
b. prokaryotes
c. euglena
d. spores

______ **6.** What kind of prokaryote can live where other organisms cannot survive?
a. archaea
b. bacteria
c. protista
d. fungi

DOMAIN BACTERIA

______ **7.** Which vitamin is produced in the human intestines by bacteria?
a. vitamin C
b. vitamin A
c. vitamin K
d. vitamin E

______ **8.** What kind of food do bacteria help us make from milk?
a. ice cream
b. milkshakes
c. whipped cream
d. yogurt

______ **9.** What kind of diseases can some kinds of bacteria cause in people?
a. colds
b. flu
c. pneumonia
d. warts

DOMAIN EUKARYA

Kingdom Protista

Match the correct description with the correct term. Write the letter in the space provided.

______ **10.** organisms that have nuclei and cell membranes

______ **11.** single- or multicellular organisms that are not fungi, plants, or animals

______ **12.** plantlike organisms in the kingdom Protista

______ **13.** animal-like organisms in the kingdom Protista

______ **14.** domain made up of all eukaryotes

a. protists
b. algae
c. eukaryotes
d. Eukarya
e. protozoans

Kingdom Fungi

Write the letter of the correct answer in the space provided.

_____ **15.** What can plants do that fungi cannot?
a. cellular respiration
b. fermentation
c. photosynthesis
d. digestion

_____ **16.** Where do fungi get their nutrients?
a. from the air
b. from the sun
c. from their surroundings
d. from energy

_____ **17.** How do fungi obtain nutrients?
a. They digest and absorb them.
b. They chew and swallow them.
c. They produce them.
d. They capture and eat them.

Kingdom Plantae

_____ **18.** What kind of eukaryotic organisms have cell walls and make their own food?
a. animals
b. plants
c. Archaea
d. fungi

_____ **19.** What must most plants be exposed to for photosynthesis to occur?
a. rainwater
b. food
c. sunlight
d. animals

_____ **20.** What do plants provide for many other organisms?
a. fungi
b. protozoa
c. sunlight
d. food and a place to live

Kingdom Animalia

______ **21.** What characteristics do most members of kingdom Animalia have?
a. They are unicellular and green.
b. They have cell walls.
c. They are multicellular and can move.
d. They perform photosynthesis.

______ **22.** What do sense organs allow animals to do?
a. to digest their food
b. to respond to their environment
c. to grow
d. to rest

______ **23.** What do animals depend on bacteria and fungi for?
a. to recycle nutrients in the environment
b. to supply chlorophyll
c. to eat
d. to use sunlight

______ **24.** Which of these is an example of a very simple animal that cannot move and has no sense organs?
a. tortoise
b. beetle
c. sponge
d. bird

Name ______________________ Class ______________ Date ______________

Skills Worksheet

Directed Reading B

Section: Sorting It All Out

1. What is classification?

____

____

WHY CLASSIFY?

______ **2.** Which of the following is NOT a question typically answered by classification?
a. What are the defining characteristics of each species?
b. What living organisms will eventually become extinct?
c. When did the characteristics of an organism evolve?
d. What are the relationships between various species?

______ **3.** Classifying living things helps scientists
a. make sense of useful animals only.
b. make sense of living things in the world.
c. make sense of geographic features.
d. make sense of the useful plants only.

HOW DO SCIENTISTS CLASSIFY ORGANISMS?

______ **4.** Taxonomy is the science of
a. stuffing and mounting animals.
b. describing, classifying, and naming organisms.
c. naming and describing rocks.
d. describing geographical features.

______ **5.** The eight-level system of classification that is similar to the one that Carolus Linnaeus developed
a. included only plants.
b. is no longer used.
c. is still used today.
d. does not include plants.

______ **6.** The more closely related living things are to each other, the more
a. characteristics they share.
b. food they share.
c. space they share.
d. water they will share.

______ **7.** Organisms are thought to be closely related when they have
a. almost no characteristics in common.
b. no characteristics in common.
c. few characteristics in common.
d. many characteristics in common.

______ **8.** Lions and house cats can both retract their claws, but brown bears and platypuses cannot. Which two are most clearly related?
a. lions and house cats.
b. lions and platypuses.
c. house cats and platypuses.
d. house cats and brown bears.

9. What Swedish botanist and physician created the first organized, modern taxonomy?

____

10. How many levels of classification do scientists use today?

____

11. Why are the platypus, brown bear, lion, and house cat thought to be related to each other?

____

____

____

12. What characteristics do the bear, lion, and house cat have that the platypus does not have?

____

____

LEVELS OF CLASSIFICATION

______ **13.** All organisms are
a. classified into one of three domains.
b. classified into one of six phyla.
c. either plants or animals.
d. either living or nonliving things.

______ **14.** Each kingdom of organisms is first divided into several
a. genera.
b. classes.
c. orders.
d. phyla.

______ **15.** The smallest, most specific classification level is

a. phylum.
b. species.
c. class.
d. order.

16. The plural form of the word *phylum* is ______________________.

17. The singular form of the word *genera* is ______________________.

18. In order from largest to smallest, what are the eight levels of classification?

____
____
____
____
____
____
____
____

SCIENTIFIC NAMES

19. No matter how many common names an organism might have, it only has one ______________________.

20. How was the naming of organisms different before Carolus Linnaeus, and how was the system difficult for scientists?

____
____
____

21. Who simplified the naming of living things by giving each species a two-part scientific name?

____

22. In the scientific name for the Asian elephant, *Elephas maximus*, the word *Elephas* indicates the animal's ______________________.

23. All genus names begin with a(n) ______________________.

24. All specific names begin with a(n) ______________________.

25. Scientific names contain information about a(n) ______________________.

26. Scientific names are usually in one of these two languages,

______________________ or ______________________.

27. In the scientific name *Tyrannosaurus rex, rex,* the Latin word for "king," is

the ______________________ name.

28. What abbreviation do scientists sometimes use when referring to *Tyrannosaurus rex*?

____

EXTINCT ORGANISMS AND LIVING ORGANISMS

29. What can scientists tell about an extinct organism in relation to a living organism with which it has many derived characteristics in common?

____

FOSSILS AND BRANCHING DIAGRAMS

30. When did the *Hypohippus* appear? Into what organism did it likely evolve?

____

____

Name ______________________ Class ______________ Date ____________

Skills Worksheet

Directed Reading B

Section: Domains and Kingdoms

1. Before the discovery of organisms like euglena, all organisms were classified as either ______________________ or ______________________.

THREE DOMAINS

______ **2.** Which characteristic is NOT true for organisms of the genus *Euglena*?
a. single celled
b. live in salt water
c. live in pond water
d. make their own food

______ **3.** What level of classification did scientists create when they realized new kingdoms could not solve larger classification problems?
a. species
b. genus
c. families
d. domains

______ **4.** A green color and the ability to make food through photosynthesis might make some people think that members of the genus *Euglena* are
a. trees.
b. herbs.
c. plants.
d. mosses.

______ **5.** Which is a characteristic that animals and members of the genus *Euglena* possess but plants do not?
a. ability to move by themselves
b. ability to make food
c. ability to take in water
d. ability to use energy

6. Scientists added the kingdom ______________________, creating a classification for organisms that were hard to classify as either fungi, plants, or animals.

7. Today, there are ______________________ domains in the classification system.

DOMAIN ARCHAEA

8. Single-celled organisms that do not have a nucleus are called

______________________.

DOMAIN BACTERIA

Identify the correct domains for the organisms described below by writing *Archaea* or *Bacteria* in the space provided.

______________________ **9.** Some of these live inside humans.

______________________ **10.** One of these causes pneumonia.

______________________ **11.** These live in places where most other organisms could not live.

______________________ **12.** Some of these can be found in hot springs in Yellowstone National Park.

______________________ **13.** One type turns milk into yogurt.

DOMAIN EUKARYA

14. All organisms whose cells have a nucleus and membrane-bound organelles are called ______________________.

15. All eukaryotes belong to the domain ______________________.

16. Members of the kingdom Protista are called ______________________.

17. Protists that have animal-like characteristics are called

______________________.

18. Protists that have plantlike characteristics are called

______________________.

19. Unlike plants, fungi do not use ______________________.

20. Unlike animals, ______________________ do not eat food.

21. How do fungi absorb nutrients from their surroundings?

____

22. Give two examples of fungi.

____

____

23. What do all members of the kingdom Plantae have in common?

____

____

____

24. In order for most plants to make their own food through photosynthesis, they must be exposed to ______________________.

25. Plants can be found growing on ______________________ and in ______________________ that light can penetrate.

26. Explain why the food that plants make is important not only to the plants themselves but to other organisms as well.

____

____

____

27. Besides making food that can be used by other organisms, plants also provide nutrients. What is another way that trees and flowering plants are used by animals?

____

____

____

28. What characteristics do most members of kingdom Animalia share?

____

____

29. Members of kingdom Animalia have specialized sense organs that allow them to respond to their ______________________.

30. Members of kingdom Animalia are commonly called ______________________.

31. Explain why animals need plants.

____

____

32. Explain how animals depend on bacteria and fungi.

____

____

STRANGE ORGANISMS

33. The kingdom Animalia includes some very simple animals, such as ______________________, that do not have sense organs and cannot move.

Name ______________________ Class ______________ Date ______________

Skills Worksheet

Vocabulary and Section Summary A

Sorting It All Out

VOCABULARY

In your own words, write a definition of the following terms in the space provided.

1. classification

____

____

2. taxonomy

____

____

SECTION SUMMARY

Read the following section summary.

- Classification groups organisms based on their shared derived characteristics.
- Classification is a tool that helps us understand the relationships between organisms.
- There are eight levels of classification.
- The scientific name of an organism has two parts.
- Branching diagrams show evolutionary relationships between extinct and living organisms.

Name ________________ Class ________________ Date ________________

Skills Worksheet

Vocabulary and Section Summary A

Domains and Kingdoms

VOCABULARY

In your own words, write a definition of the following terms in the space provided.

1. Archaea

2. Bacteria

3. Eukarya

4. Protista

5. Fungi

6. Plantae

7. Animalia

SECTION SUMMARY

Read the following section summary.

- Most biologists recognize three domains: Archaea, Bacteria, and Eukarya.
- As scientists discover new organisms, classification systems are changed to include the characteristics of those new organisms.
- Archaea can live in extreme environments. Bacteria live almost everywhere else. All prokaryotes are members of the domain Archaea or the domain Bacteria.
- Domain Eukarya is made up of four kingdoms: Protista, Fungi, Plantae, and Animalia. All members of Eukarya are eukaryotes.

Name ______________________ Class ______________ Date ______________

Skills Worksheet

Vocabulary and Section Summary B

Sorting It All Out

VOCABULARY

After you finish reading the section, try this puzzle! Use the clues given to fill in the blanks below. Then, copy the numbered letters into the corresponding boxes to reveal the levels of classification.

1. the science of describing, classifying, and naming organisms

___	___	___	___	___	___	___	___
				7			2

2. classifying organisms by their characteristics

___	___	___	___	___	___	___	___	___	___	___
					4	10				

3. putting things in orderly groups based on their characteristics

___	___	___	___	___	___	___	___	___	___	___	___	___	___
	3					11						13	

4. shows which characteristics organisms share and when they evolved

___	___	___	___	___	___	___	___	___	___	___	___	___	___	___	___
	14				1				15			5			

5. characteristics that two or more kinds of organisms share with their most recent common ancestor

___	___	___	___	___	___	___	___	___	___	___	___	___
				6					12			

___	___	___	___	___	___	___	___	___	___	___	___	___	___	___
9										8				

P				U		■				U		■					
	1	2	3		4		5	6	7		8		9	3	10	8	8

						■						■		P					
11	10	4	12	3	2		13	14	15	6	14		8		6	9	12	6	8

						■	K						
15	13	4	10	12	7	51		12	7	5	15	13	4

Organize the levels of classification from most to least general below.

a. ______________________________

b. ______________________________

c. ______________________________

d. ______________________________

e. ______________________________

f. ______________________________

g. ______________________________

h. ______________________________

SECTION SUMMARY

Read the following section summary.

- Classification groups organisms based on their shared derived characteristics.
- Classification is a tool that helps us understand the relationships between organisms.
- There are eight levels of classification.
- The scientific name of an organism has two parts.
- Branching diagrams show evolutionary relationships between extinct and living organisms.

Name ______________________ Class ______________ Date ______________

Skills Worksheet

Vocabulary and Section Summary B

Domains and Kingdoms

VOCABULARY

After you finish reading the section, try this puzzle! In the space provided, write the term described. Then, find those words in the word search puzzle on the next page. Terms can be hidden in the puzzle vertically, horizontally, diagonally, or backward.

For each of the following descriptions, write the kingdom or domain of the organisms being described.

______________________ **1.** single-celled organisms without nuclei, such as *Escherichia coli*, which live in the human body

______________________ **2.** multicellular, eukaryotic organisms that are usually green and make sugar through photosynthesis

______________________ **3.** prokaryotes that were first discovered living in extreme environments where other organisms could not survive

______________________ **4.** multicellular organisms whose cells have nuclei but do not have cell walls

______________________ **5.** multicellular organisms that have cells containing nuclei and that absorb nutrients from their surroundings after breaking them down with digestive juices

______________________ **6.** single-celled or simple multicellular eukaryotic organisms

______________________ **7.** organisms whose cells have a nucleus and membrane-bound organelles

D	I	C	H	O	T	A	Y	R	A	K	U	E	Y	I
Y	A	L	S	Y	A	R	Z	A	U	R	Q	W	F	J
G	F	A	S	P	X	E	S	K	P	P	X	B	I	U
P	K	S	O	H	O	X	A	N	S	S	K	A	F	Y
E	E	S	F	U	N	G	I	I	N	W	I	C	A	Y
C	G	I	Y	G	O	F	T	P	L	A	N	T	A	E
P	D	F	H	X	M	K	F	L	K	A	S	E	C	Q
W	M	I	G	H	Y	N	F	O	Z	I	M	R	F	P
A	R	C	H	A	E	A	A	C	T	E	R	I	A	M
P	G	A	R	L	R	W	S	O	F	S	R	A	N	I
V	U	T	J	X	I	F	R	L	B	V	N	Y	F	A
J	W	I	S	T	W	P	U	O	A	F	N	L	F	D
D	S	O	Y	G	E	O	I	X	Y	R	A	C	K	K
S	F	N	E	K	D	W	D	X	Q	H	Z	T	S	C

SECTION SUMMARY

Read the following section summary.

- Most biologists recognize three domains: Archaea, Bacteria, and Eukarya.
- As scientists discover new organisms, classification systems are changed to include the characteristics of those new organisms.
- Archaea can live in extreme environments. Bacteria live almost everywhere else. All prokaryotes are members of the domain Archaea or the domain Bacteria.
- Domain Eukarya is made up of four kingdoms: Protista, Fungi, Plantae, and Animalia. All members of Eukarya are eukaryotes.

Name ______________________ Class ______________ Date ______________

Skills Worksheet

Reinforcement

Keys to the Kingdom

Complete this worksheet after you finish reading the section "Domains and Kingdoms."

Patty dropped her notes while she was studying the four kingdoms of living things in the domain Eukarya, and now she isn't sure which facts belong to which kingdom. Each of the four boxes on the next page is labeled with the name of one of the kingdoms. Help Patty out by listing the facts, descriptions, and examples from her notes below in the appropriate boxes.

- Have specialized sense organs
- Break down material outside their body and then absorb the nutrients
- Usually green
- Complex organisms with no cell walls
- Most are single-celled or simple multicellular organisms
- Make food through photosynthesis
- Protozoans
- Algae
- Molds
- Do not perform photosynthesis
- Evolved from bacteria about 2 billion years ago
- Mushrooms
- Use the sun's energy to make sugar
- Giant sequoias

Protista	Fungi

Plantae	Animalia

Name ______________________________ Class ______________ Date ____________

Skills Worksheet

Critical Thinking

A Breach on Planet Biome

Dear Professor Noah Itall,

I am the Head Counselor of the Noid Clan on Planet Biome. I am writing because I need your help.

The Druffs are a neighboring clan. They have recently become our close allies. The other day, the clan leaders, Dan Druff and Anne Noid, were discussing the best way to unite the clans. Anne noticed a plant nearby that we call "itchyskinda," a plant that makes your skin develop a red rash. Anne tried to warn Dan about the itchyskinda, but Dan did not understand this name. Much to his dismay, Dan rubbed against the plant and developed an irritating rash.

Our clans have similar languages, but some words are different. Each clan has a different name for plants and animals. It has been hard to share our knowledge of different species. We have considered using half the species names from each clan to identify plants and animals. This way, each clan would need to learn new names for only half of the plants and animals.

I decided to write to you for advice. What do you think of the proposed solution? I'm afraid that if I do not solve the problem soon, our allies might soon become our enemies.

Please write soon,

Perry Noid

STATE THE PROBLEM

1. In your own words, state the main problem between the Noid and Druff clans.

____

____

____

DETERMINING CAUSE AND EFFECT

2. What is the cause of the main problem?

____

____

____

____

____

3. What are the effects of the problem between the clans?

____

____

____

____

EVALUATING INFORMATION

4. Do you think it would be a good idea to use half the species names from each clan to identify plants and animals? Explain your answer.

____

____

____

____

____

APPLYING CONCEPTS

5. How do you think Perry Noid should solve this problem? Recommend a solution.

____

____

____

____

____

____

Name ______________________ Class ______________ Date ______________

Activity

SciLinks Activity

KINGDOMS

Go to www.scilinks.org. To find links related to kingdoms of organisms, type in the keyword HY71397. Then, use the links to answer the following questions about kingdoms.

Internet Resources

For a variety of links related to this chapter, go to www.scilinks.org

Topic: Kingdoms

SciLinks code: HY71397

1. Use the space below to draw a member of kingdom Fungi in a habitat that provides nutrients to the fungus.

2. Draw a member of kingdom Plantae in its natural habitat.

3. Draw a picture that shows how a member of the kingdom of Animalia depends on a member of the kingdom Plantae.

Name ______________ Class ______________ Date ______________

Skills Worksheet

Section Review

Sorting It All Out

USING VOCABULARY

1. Write an original definition for *classification* and *taxonomy.*

____

____

UNDERSTANDING CONCEPTS

2. Analyzing Why do scientists use scientific names for organisms?

____

____

3. Listing What are the eight levels of classification?

____

____

INTERPRETING GRAPHICS

Use the branching diagram below to answer the next two questions.

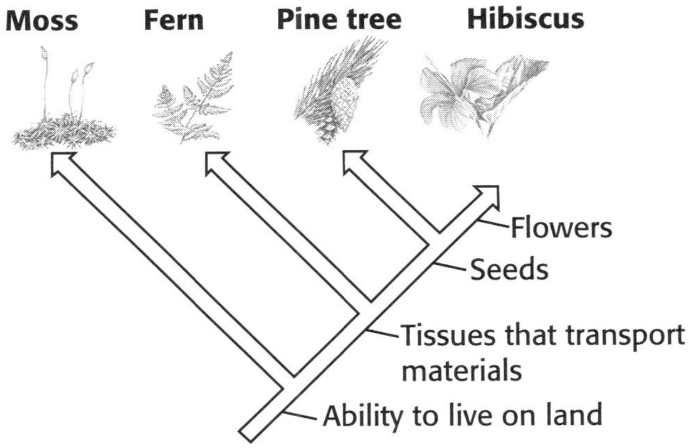

4. Identifying Which kind of organism evolved earliest? Which kind of organism evolved most recently?

____

5. Inferring Which organisms have tissues that transport materials?

____

____

CRITICAL THINKING

6. Analyzing Processes You have found a fossil of an organism. The organism has characteristics that have never been described. What would you have to do to identify this fossil?

____

____

____

____

7. Making Inferences What is the difference between organisms that share many derived charateristics and organisms that do not?

____

____

____

____

8. Applying Concepts There is an organism halfway up a branch in a branching diagram, and a different organism at the tip of the branch. What can you infer about these organisms?

____

____

____

____

9. Making Inferences In branching diagrams, what can you infer about the organisms at the tips of the branches that do not reach the tops of the diagrams?

____

____

____

____

Name ______________________ Class ______________ Date ____________

Skills Worksheet

Section Review

Domains and Kingdoms

USING VOCABULARY

1. Write an original definition for *Archaea* and *Bacteria*.

____

____

UNDERSTANDING CONCEPTS

2. Identifying Describe one characteristic of the domain Eukarya and one characteristic of each kingdom in the domain Eukarya.

____

____

CRITICAL THINKING

3. Applying Concepts What do all of the organisms from the three domains have in common?

____

____

4. Making Inferences A branching diagram shows an unusual group of organisms on a branch between birds and mammals. What could you tell about when this unusual group of organisms evolved?

____

____

____

5. Applying Concepts You have discovered a new prokaryote. It lives deep within the crust of Earth, which is an extreme environment that has little air or food. Into which domain would you classify this organism? Explain.

____

____

____

____

____

MATH SKILLS

6. Making Calculations If a certain bacterium divides every 30 min, when will there be more than 1,000 bacteria? Show your work below.

CHALLENGE

7. Identifying Relationships Very hot water gushes out of formations on the sea floor called black smokers. Scientists have discovered many new kinds of organisms around black smokers. Why are scientists using a classification system to group these organisms?

____

____

____

____

8. Applying Concepts To get nutrients, the Venus' flytrap uses photosynthesis and traps and digests insects. Its cells have cell walls. Into which kingdom would you place this organism? What makes this organism unusual in this kingdom?

____

____

____

____

Name ______________________ Class ______________ Date ____________

Skills Worksheet

Chapter Review

USING VOCABULARY

______ **1. Academic Vocabulary** In the sentence "Scientists classify organisms based on their shared derived characteristics," what does the word *derived* mean?

a. caused by
b. inherited from
c. rare
d. common

Complete each of the following sentences by choosing the correct term from the word bank.

Archaea
eukaryotes
prokaryotes
Animalia

2. Prokaryotes that live in extreme environments are in the domain

______________________.

3. Complex multicellular organisms that can usually move around and respond to their environment are in the kingdom ______________________.

4. Although ______________________ are very similar, they are divided into two domains in part because they can live in extremely different environments.

UNDERSTANDING CONCEPTS

Multiple Choice

______ **5.** How do scientists classify organisms?

a. by grouping the organisms by their characteristics
b. by giving the organisms many common names
c. by deciding whether the organisms are useful
d. by using only existing categories of classification

______ **6.** The scientific name for the European white water lily is *Nymphaea alba*. To which genus does this plant belong?

a. *Nymphaea*
b. *alba*
c. water lily
d. alba lily

Short Answer

7. Identifying Why is the use of scientific names important in biology?

____

____

8. Evaluating What kind of evidence is used by modern taxonomists to classify organisms based on evolutionary relationships?

____

____

9. Demonstrating Is a bacterium a type of eukaryote? Explain your answer.

____

____

INTERPRETING GRAPHICS

Use the branching diagram of selected primates below to answer the next three questions.

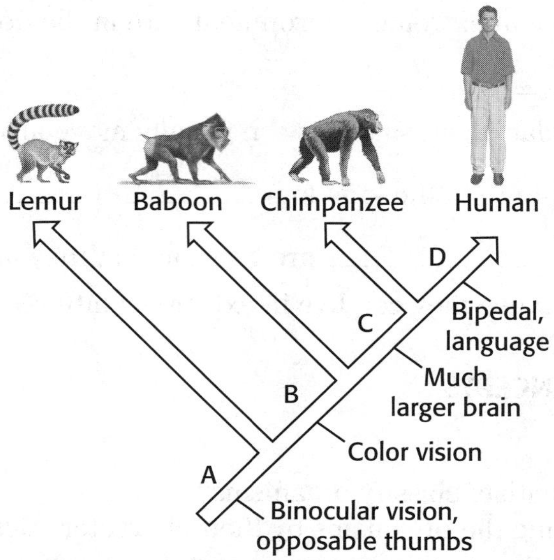

10. Identifying Which primate is the closest relative to the common ancestor of all primates?

____

11. Evaluating Which primate shares the most characteristics with humans?

____

12. Demonstrating Do both lemurs and humans have the characteristics listed at point D? Explain your answer.

____

____

____

WRITING SKILLS

13. Writing Persuasively In the past, scientists classified organisms as either plants or animals. Why doesn't that classification system work? Make sure to use examples to support your position.

CRITICAL THINKING

14. Concept Mapping Use the following terms to create a concept map: *domain, kingdom, fern, lizard, Animalia, Fungi, algae, Protista, Plantae,* and *mushroom.*

15. Analyzing Methods Explain how the levels of classification depend on the similarities and differences between organisms.

____

____

____

____

16. Identifying Relationships What characteristic do all of the members of the four kingdoms in the domain Eukarya share?

____

17. Applying Concepts You have discovered the fossil of an organism that has a tail, gills, and legs. What steps would you take to figure out where this fossil would fit on a branching diagram that contains, fish, amphibians, reptiles, birds and mammals?

____

____

____

____

____

18. Making Inferences Sometimes, the branches in a branching diagram stop at dead ends. These dead ends may not reach the top of the diagram. What can you tell about the organisms shown at the end of these branches? Explain.

____

____

____

____

19. Identifying Relationships All organisms are made up of cells. Prokaryotes are made up of single cells but are divided into two domains. Describe a way in which you could determine the domain in which to classify a prokaryote.

____

____

____

____

20. Expressing Opinions Sam has discovered a new organism that has some strange features. He has decided to create a new phylum for this organism. Do you think that this is the correct way to classify this organism? Explain your position.

____

____

____

____

____

INTERPRETING GRAPHICS

Use the photo in your book to answer the next three questions.

21. Evaluating Data Describe two basic characteristics of the organism in the photo.

____

____

22. Identifying Relationships Using the characteristics of the organism in the photo, can you decide whether this organism is a fish, a reptile, a bird, an animal, or a plant? Explain.

____

____

____

____

23. Expressing Opinions What other kinds of information could help you decide how to classify the organism in the photo?

____

____

____

____

MATH SKILLS

24. Making Conversions Many prokaryotes live near deep-sea hydrothermal vents. The water coming out of the vents can be hotter than 536°F. Just a few meters away from the vent, the water temperature can be as cool as 35.6°F. Calculate the range of temperatures (in degrees Celsius) in which the prokaryotes live. (Hint: Convert both temperatures into degrees Celsius first.) Show your work below.

CHALLENGE

25. Identifying Relationships The kingdom Protista is made up of single-celled organisms and multicellular organisms. These organisms are classified into the kingdom Protista because they do not fit into the other kingdoms. Explain why this characteristic makes kingdom Protista unusual when compared to other kingdoms.

____

____

____

____

____

____

____

____

Assessment

Chapter Pretest

Teacher Notes and Answer Key

The Pretest questions are designed to help you determine the prior knowledge of your students. Some questions test whether students have mastered the background knowledge they need to understand the content you are about to teach. Other questions test your students' prior knowledge of the content you are about to teach. Use the Pretest with the Test Doctors and diagnostic teaching tips in these notes pages to help you tailor your instruction to your students' specific needs.

QUESTION NUMBER	CORRECT ANSWER	STANDARD
1	*D*	7.3.e
2	*B*	7.3.a
3	*C*	7.3.b
4	*D*	7.3.b
5	*B*	7.3.c
6	*C*	7.3.d
7	*C*	7.3.d
8	*D*	7.3.d
9	*D*	7.1.a

TEST DOCTOR

The following Pretest questions have been diagnosed by the Test Doctor. Find out what might be causing your students' "ailing" answers. Each Test Doctor is followed by a diagnostic teaching tip to help you address students' learning needs.

Question 1 *asks students to identify the factors that can lead to extinction.*

A Incorrect. Competition for shelter and space can lead to extinction.

B Incorrect. New predators could cause extinction if their prey cannot adapt

C Incorrect. Forest fires can cause massive changes to the ecosystem and put many organisms at risk of extinction.

D Correct. New food sources would allow increases in population sizes and are not likely to lead directly to the extinction of a species.

Diagnostic Teaching Tip: Students who have difficulty with this question should discuss the many types of environmental changes that can occur in an ecosystem and how plants and animals would have to adapt to continue to survive in the new environment. Discussion should cover scenarios such as natural disasters, competition from new species, and loss of a food source. Stress that organisms do not make conscious choices to adapt to an environment. Successful organisms are those organisms that happen to have those characteristics that allow them greater success than their peers.

Question 2 *asks students to identify which characteristic gives the greatest survival advantage to a species in an ecosystem with environmental change.*

A Incorrect. Species with little genetic variety are vulnerable to changes.
B Correct. Species with large numbers of different alleles are more likely to have the adaptations for changes in the environment.
C Incorrect. Species whose members all have the same type of alleles are very vulnerable to changes and extinction.
D Incorrect. Species with little variety in traits have little genetic variation and are less likely to be able to adapt to changes.

Diagnostic Teaching Tip: Students who have problems with this question should review how increased genetic variability gives a species a greater probability of being able to adapt to changes. Work as a class to create an example that compares two species, one with several alleles for resistance to a disease and the other with only one allele. Have the class decide how the populations of each species would be affected by the disease.

Question 3 *asks students to identify factors that affect a population's size.*

A Incorrect. Competition for limited resources will lead to a decrease in population size as some of its members leave or die.
B Incorrect. Disease is a limiting factor for a species, since it may result in the death of some members.
C Correct. Overproduction is a component of natural selection and is more likely to result in at least a temporary increase in the population.
D Incorrect. Predation will result in the death of some members of the species and a consequent reduction in population size.

Diagnostic Teaching Tip: Students who have difficulty with this question should be given an example of an ecosystem with a limited food source, the species of interest, a competing species, and a predator. Show how predation and competition for food limit the species of interest. Conclude by discussing the results a major disease would have on the population.

Question 4 *asks students to know Darwin's theory of natural selection.*

A Incorrect. Darwin knew that organisms inherit traits, but not how they inherit traits.
B Incorrect. Darwin found that predation and competition meant that only a few offspring would survive to adulthood.
C Incorrect. Darwin's research led him to conclude that some organisms are better adapted to their environments than others.
D Correct. Darwin found that parents who are successful in their environment are more likely to live and pass those adaptive traits on to their offspring.

Diagnostic Teaching Tip: Students who have trouble with this question should study the four parts of natural selection as proposed by Darwin. Organize the class in four groups, and assign each one of the parts. They should research an example that demonstrates their part. Have groups clearly connect genetic variation and successful adaptations being passed to future generations. When groups are finished, have them present their findings to the class.

Question 5 *asks students to determine which area (layer) of fossil recovery has the most recent organisms.*

A Incorrect. Area A, although near a lake, is not in the uppermost layer of rock where fossils were found.

B Correct. Area B is in the uppermost layer of rock where fossils were found. This is where the fossils of the most recently deceased organisms would be found.

C Incorrect. Area C is where many fossils were found, but it is not the uppermost layer of rock where fossils were found.

D Incorrect. Area D is the lowest layer of rock where fossils were found. These fossils would be the oldest of the fossils in this diagram.

Diagnostic Teaching Tip: Students who have difficulty answering this question might be confused by where the groups of fossils were located geographically or by the number of fossils that were found. Remind students that, when determining which fossil is the most recent in terms of evolution, one needs to know in which layer of rock the fossil was found. Fossils found closest to the surface of Earth are most likely to be the fossils of the most recently deceased organisms, regardless of where they are geographically in an area or how many fossils were found.

Question 6 *asks students to know how scientists classify organisms.*

A Incorrect. Many different species of organisms live in the same habitats.

B Incorrect. Many different species occur in similar time periods.

C Correct. The most useful way to categorize organisms is by their similar characteristics, such as their physical and genetic characteristics.

D Incorrect. Length of life span is not a useful distinction. Many species have similar life spans.

Diagnostic Teaching Tip: Students who have difficulty answering this question should study the different taxonomic levels and how organisms are grouped by similarities in their characteristics within each level. Show the class a branching diagram for the phylum Chordata. This phylum contains only organisms with backbones. In the next category, class, animals are further differentiated when scientists classify them as mammals. Categories are not related to the size of the animal or when it lived. Be sure to emphasize Section 1, "Sorting It All Out," in Chapter 11, "Classification." Students must be able to classify living organisms in order to master standard 7.3.d.

Question 7 *asks students to indicate what taxonomic level is used as the first part of an organism's scientific name*

A Incorrect. Domain is the highest level of heirarchical classification. It includes many kinds of organisms.

B Incorrect. The family level is not the first part of an organism's scientific name. It includes many genera.

C Correct. The genus level of classification is directly above the species level. It is used as the first name of the two-part scientific name of an organism.

D Incorrect. Kingdom is a broad category of many organisms. It is not the first part of an organism's scientific name.

Diagnostic Teaching Tip: Students who have difficulty answering this question might benefit from seeing examples of two-part scientific names as developed by Linnaeus. Show students how two species in the same genus share the genus trait but not some species' traits. Be sure to emphasize Section 1, "Sorting It All Out," in Chapter 11, "Classification." Students must be able to classify living groups of organisms by shared derived characteristics in order to master standard 7.3.d.

Question 8 *asks students to identify what level of scientific taxonomy organisms would have to share to be able to mate and produce offspring.*

A **Incorrect.** Organisms at the class level are not related closely enough to produce fertile offspring.

B **Correct.** Organisms at the family level are not related closely enough to produce fertile offspring.

C **Incorrect.** Organisms at the genus level are not related closely enough to produce fertile offspring.

D **Correct.** According to the biological species definition, only organisms at the species level are related closely enough to produce fertile offspring.

Diagnostic Teaching Tip: Students who have trouble answering this question correctly might benefit from a discussion of how mating with other organisms that are similar to themselves allows that particular species to survive. Ask students to consider what would happen if organisms mated between species. Help them understand that the original species may become extinct as a consequence. Therefore, many mechanisms that prevent mating between different species of organims have arisen within different species. Be sure to emphasize Section 1, "Sorting It All Out," in Chapter 11, "Classification." Students must be able to classify living groups of organisms by shared derived characteristics in order to master standard 7.3.d.

Question 9 *asks students to indicate what scientists have found that all organisms have in common.*

A **Incorrect.** Organisms such as animals do not have cell walls around their cells.

B **Incorrect.** Protists include a wide variety of organisms and cannot be the ancestors of all the organisms that have ever evolved.

C **Incorrect.** All organisms that have the same sets of genes also have the same characteristics, which would not account for the diversity of organisms.

D **Correct.** All organisms are made of one or more cells that have similar mechanisms to carry out similar basic life functions.

Diagnostic Teaching Tip: Students who struggle with this question should benefit from an introduction to the three domains. The domains Archaea and Bacteria consist of single-celled organisms with no nucleus or organelles. The domain Eukarya consists of organisms whose cells have a nucleus and organelles. Be sure to emphasize Section 2, "Domains and Kingdoms," in Chapter 11, "Classification." Students must know that cells function similarly in all living organisms, in spite of diversity in the domains, in order to master standard 7.1.a.

Name ______________________ Class ______________ Date ____________

Assessment

Chapter Pretest

______ **1.** Which of the following changes to an environment would be least likely to lead to the extinction of a species?
A fewer nesting sites
B new predators
C forest fires
D new food sources

______ **2.** Which of the species with the following traits will have the greatest survival advantage when there is a change in the environment?
A those with only a few kinds of genes
B those with many different alleles
C those with the same type of alleles
D those with little variety in traits

______ **3.** Which of the following factors would likely have the least impact on the growth of populations?
A competition
B disease
C overproduction
D predation

______ **4.** Darwin developed the theory of natural selection to explain the variations and adaptations among organisms. Which of the following was an important part of Darwin's theory?
A Many organisms have exactly the same set of traits.
B All offspring will survive to be adults.
C All organisms are well adapted to their environments.
D Offspring inherit adaptations that help their parents succeed.

_____ **5.** A scientist is studying the fossils that were recovered from an area to determine the most recent ancestor of a species known to live there. Where are the most recent fossils likely to be found?

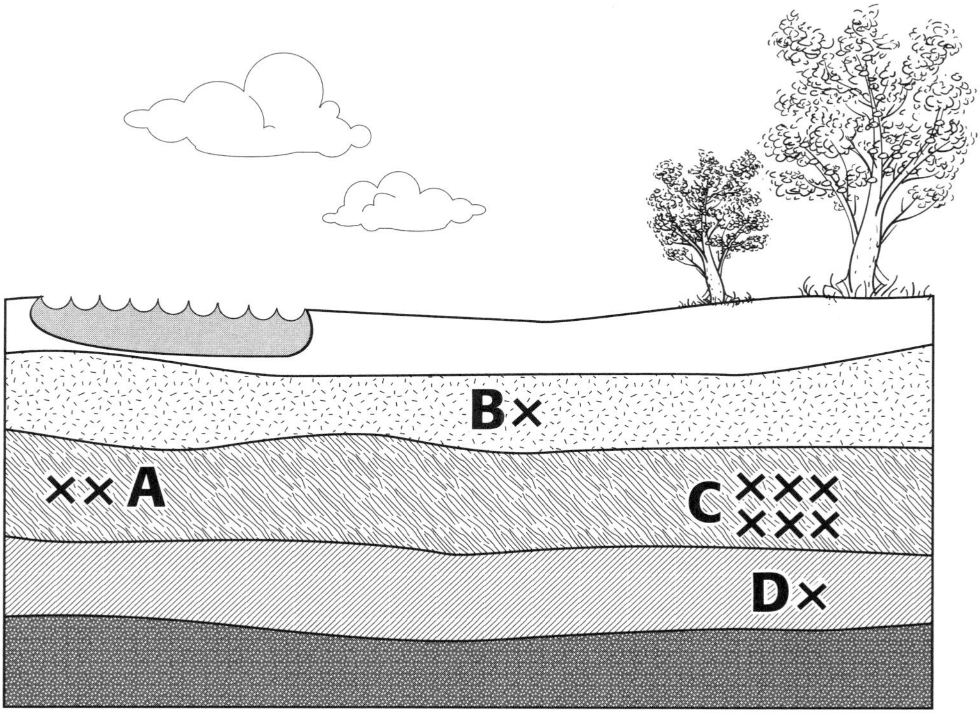

A area A
B area B
C area C
D area D

_____ **6.** Scientists classify plants and animals so that they can make some sense and order of the many different types of organisms found on Earth. What do scientists base their grouping of living things on?
A living in similar habitats
B living in similar time periods
C having similar characteristics
D having similar life spans

_____ **7.** Scientific names are derived from Latin or Greek words and describe the specific organism. Each scientific name has two parts. The second part of the name is based on the species. What is the first part of the name based on?
A domain
B family
C genus
D kingdom

______ **8.** The major levels of classification for biological organisms are in an ordered hierarchy: domain, kingdom, phylum, class, order, family, genus, species. If you wanted to breed your dog with another dog to produce puppies, what level of classification must both animals be from in order for them to have fertile offspring?

A class
B family
C genus
D species

______ **9.** Developing classifications and examining fossil evidence have allowed scientists to learn many things about the evolution of organisms. What have scientists found that all organisms have in common?

A They all have cell walls around their cells.
B They evolved from protists.
C They all have the same sets of genes.
D They are all made of one or more cells.

Assessment

Section Quiz

Section: Sorting It All Out

Match the correct definition with the correct term. Write the letter in the space provided.

_____ **1.** the science of describing, classifying, and naming organisms

_____ **2.** the division of organisms into groups, or classes, based on specific characteristics

a. classification

b. taxonomy

Write the letter of the correct answer in the space provided.

_____ **3.** Today, what system based on shared characteristics is used to classify organisms?
- **a.** sets of encyclopedias
- **b.** the eight-level system
- **c.** notes and photographs
- **d.** drawings and graphs

_____ **4.** In order, what are the three levels of classification that are used in addition to domain, kingdom, family, genus, and species?
- **a.** phylum, order, class
- **b.** class, order, phylum
- **c.** phylum, class, order
- **d.** class, order, genera

_____ **5.** The scientific name for an organism comes from its
- **a.** main characteristic.
- **b.** kingdom and phylum.
- **c.** order and class.
- **d.** genus and specific name.

_____ **6.** Which of the following does a branching diagram show?
- **a.** when an organism evolved
- **b.** when an organism will become extinct
- **c.** how an organism evolved
- **d.** why an organism became extinct

Name ______________________ Class ______________ Date ______________

Assessment

Section Quiz

Section: Domains and Kingdoms

Write the letter of the correct answer in the space provided.

______ **1.** For hundreds of years, how were all living things classified?
- **a.** as living or nonliving
- **b.** as plants or animals
- **c.** as water creatures or land creatures
- **d.** as large or small

______ **2.** What happened that made it necessary for scientists to adopt the classification system?
- **a.** They discovered new organisms.
- **b.** They learned to write clearly.
- **c.** They started using photography.
- **d.** They began classifying in English.

______ **3.** How many domains are recognized today?
- **a.** four
- **b.** five
- **c.** three
- **d.** ten

Match the correct definition with the correct term. Write the letter in the space provided.

______ **4.** a domain made up of prokaryotes that live in soil, water, and in the human body

______ **5.** a kingdom made up of complex, multicellular organisms that lack cell walls, can usually move around, and respond quickly to their environment

______ **6.** a kingdom of complex, multicellular organisms that are usually green, have cell walls, and use the sun's energy to make sugar by photosynthesis

______ **7.** a kingdom of mostly single-celled or simple multicellular organisms

______ **8.** a kingdom of nongreen, eukaryotic organisms that do not move and that reproduce by spores

______ **9.** a domain made up of prokaryotes that can live in extreme environments

- **a.** Protista
- **b.** Fungi
- **c.** Archaea
- **d.** Animalia
- **e.** Plantae
- **f.** Bacteria

Name ______________________ Class ______________ Date ______________

Assessment

Chapter Test A

Classification

MULTIPLE CHOICE

Write the letter of the correct answer in the space provided.

______ **1.** What do scientists look at to classify living things?
a. their age
b. their likes and dislikes
c. their characteristics
d. their diseases

______ **2.** What is the science of taxonomy?
a. naming plants and animals
b. describing, classifying, and naming living things
c. measuring living things
d. taking pictures of living things

______ **3.** In the past, what two groups did scientists use to classify all living things?
a. living and nonliving
b. tall and short
c. plant and animal
d. large and small

______ **4.** How many levels of classification do scientists use today?
a. four
b. five
c. six
d. eight

______ **5.** How many domains do scientists use to classify living things today?
a. five
b. seven
c. three
d. four

______ **6.** Branching diagrams show when an organism
a. died.
b. evolved.
c. ate.
d. migrated.

______ **7.** What makes up a living thing's scientific name?
a. its genus and specific name
b. its kingdom and phylum
c. its class and order
d. its family and genus

______ **8.** What is the scientific name for an Asian elephant?
a. *Elephas asian*
b. *Elephas maximus*
c. *Elephas biggus*
d. *Elephas rex*

______ **9.** What do we call simple, multicellular living things?
a. protists
b. bacteria
c. fungi
d. archaea

______ **10.** How do fungi take in and use nutrients from their surroundings?
a. They capture and kill them.
b. They chop and swallow them.
c. They absorb and chew them.
d. They absorb and digest them.

______ **11.** What must a plant be exposed to for photosynthesis to occur?
a. cold water
b. food
c. sunlight
d. energy

MATCHING

Match the correct description with the correct term. Write the letter in the space provided.

______ **12.** They are found where other living things cannot survive.

______ **13.** They can live in soil, water, or in the human body.

a. bacteria
b. archaea

Match the correct description with the correct term. Write the letter in the space provided.

______ **14.** kingdom made up of eukaryotes with cell walls that perform photosynthesis and vary greatly in size

______ **15.** kingdom with organisms that can respond quickly to changes in their environment

______ **16.** domain made up of all eukaryotes

______ **17.** kingdom that includes protists

a. Eukarya

b. Plantae

c. Animalia

d. Protista

FILL-IN-THE-BLANK

Use the terms from the following list to complete the sentences below.

domains	classes	species
genera	orders	families

18. Families are broken down into ______________________.

19. Kingdoms are sorted into phyla, and phyla into

______________________.

20. The largest, most general groups for classifying organisms

are ______________________.

21. Classes include one or more ______________________.

22. Orders are separated into ______________________.

23. Genera are sorted into ______________________.

Name ______________________________ Class ______________ Date ______________

Assessment

Chapter Test B

Classification

MULTIPLE CHOICE

Write the letter of the correct answer in the space provided.

______ **1.** The division of organisms into groups, or classes, based on characteristics is
a. taxonomy.
b. classification.
c. life science.
d. biology.

______ **2.** The science of describing, classifying, and naming organisms is
a. taxonomy.
b. classification.
c. life science.
d. organization.

______ **3.** Carolus Linnaeus is known for
a. founding the science of taxonomy.
b. discovering retractable claws.
c. identifying the characteristics of rare species.
d. discovering *Tyrannosaurus rex*.

______ **4.** The eight levels of classification, from general to specific, are
a. domain, kingdom, class, order, phylum, family, genus, species.
b. domain, kingdom, phylum, class, family, order, genus, species.
c. domain, kingdom, phylum, class, order, family, genus, species.
d. domain, kingdom, class, phylum, order, family, genus, species.

______ **5.** A pine tree is a member of the kingdom
a. Animalia.
b. Fungi.
c. Protista.
d. Plantae.

______ **6.** For hundreds of years, people thought all living things belonged to one of two groups. What were the two groups?
a. living and nonliving
b. animals and flowers
c. animals and plants
d. microscopic and nonmicroscopic

______ **7.** What is shown in a branching diagram?
a. characteristics organisms share and when organisms evolved
b. why organisms evolved
c. when living organisms will become extinct
d. how organisms became extinct

______ **8.** Euglena are members of the kingdom
a. Protista.
b. Plantae.
c. Animalia.
d. Fungi.

______ **9.** The organisms in what kingdom usually move by themselves and have specialized sense organs that allow them to respond to their environment?
a. Fungi
b. Plantae
c. Animalia
d. Protista

______ **10.** The scientific name of an organism comes from its
a. kingdom and phylum.
b. class and genus.
c. kingdom, phylum, and class.
d. genus and specific name.

______ **11.** The scientific name for the common house cat is *Felis domesticus*. What is its specific name?
a. *Felis*
b. *domesticus*
c. house cat
d. feline

______ **12.** What do scientists do when they discover organisms that do not fit into any of the four kingdoms?
a. destroy the newly discovered organisms
b. change the newly discovered organisms so they will fit one of the four kingdoms
c. create new categories
d. leave the newly discovered organisms alone

______ **13.** An example of a simple animal is a(n)
a. fern.
b. mold.
c. insect.
d. sponge.

______ **14.** Members of the kingdom Animalia depend on bacteria and fungi because bacteria and fungi
a. do not perform photosynthesis.
b. recycle nutrients in dead organisms.
c. use sunlight to produce sugar.
d. are useful for animal habitats.

______ **15.** What kingdom does slime mold belong to?
a. Protista
b. Fungi
c. Plantae
d. Archaea

MATCHING

Match the correct description with the correct term. Write the letter in the space provided.

______ **16.** This domain includes prokaryotes that live in extreme environments.

______ **17.** This domain includes prokaryotes that live in soil, water, and in humans.

______ **18.** This kingdom includes mostly single-celled or simple multicellular organisms.

______ **19.** This kingdom includes nongreen, nonmoving eukaryotic organisms that break down substances outside their bodies and absorb the nutrients.

______ **20.** This kingdom includes complex, multicellular organisms that lack cell walls, can usually move, and respond to their environment.

______ **21.** This kingdom includes complex, multicellular organisms that are usually green, have cell walls, and make sugar by photosynthesis.

______ **22.** This domain is made up of all eukaryotes.

a. Plantae
b. Fungi
c. Bacteria
d. Protista
e. Animalia
f. Archaea
g. Eukarya

MULTIPLE CHOICE

Use the following figure to answer questions 23 through 25. Write the letter of the correct answer in the space provided.

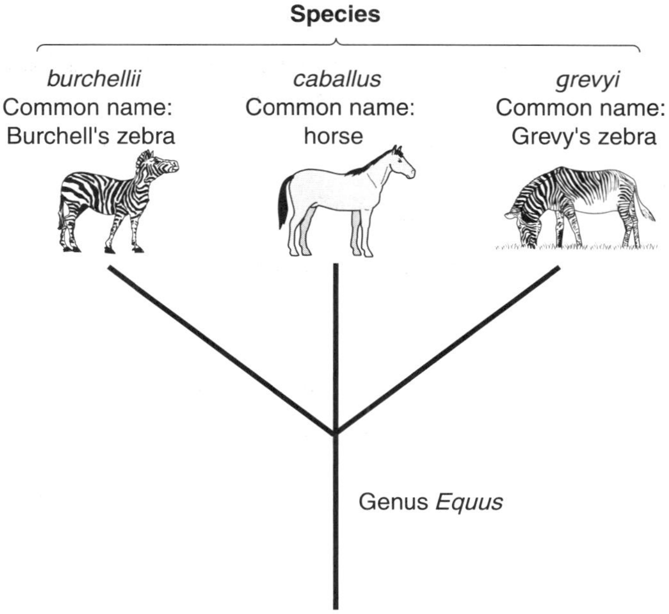

______ **23.** What is the scientific name for a Burchell's zebra?

a. *Equus caballus*
b. *Equus grevyi*
c. *Equus burchellii*
d. *Caballus zebra*

______ **24.** What is the scientific name for a horse?

a. *Equus caballus*
b. *Equus grevyi*
c. *Equus burchellii*
d. *Equus horse*

______ **25.** What is the scientific name for a Grevy's zebra?

a. *Equus caballus*
b. *Equus zebra*
c. *Equus grevyi*
d. *Equus burchellii*

Name ______________________________ Class ______________ Date ____________

Assessment

Chapter Test C

Classification

USING KEY TERMS

Use the terms from the following list to complete the sentences below. Each term may be used only once. Some terms may not be used.

classification	taxonomy	scientific name
branching diagram	Archaea	Bacteria
Protista	Animalia	Plantae

1. The science of describing, naming, and classifying organisms is

________________________.

2. Using a(n) ________________________, one can determine when an organism evolved and which characteristics organisms share.

3. Members of kingdom ________________________ are single-celled or simple multicellular organisms.

4. A giant sequoia belongs to kingdom ________________________.

5. Members of kingdom ________________________ are multicellular, are often able to move about, and respond to their environment.

6. Members of the domain ________________________ are prokaryotes that can live in extreme environments.

UNDERSTANDING KEY IDEAS

Write the letter of the correct answer in the space provided.

______ **7.** Scientists classify organisms based on their
- **a.** likes and dislikes.
- **b.** numbers in the wild.
- **c.** age.
- **d.** characteristics.

______ **8.** What do scientists use to refer to organisms because common names can create confusion?
- **a.** common names
- **b.** nicknames
- **c.** scientific names
- **d.** first names

______ **9.** The three most general levels of classification, from general to specific, are
- **a.** kingdom, family, and class.
- **b.** domain, kingdom, and phylum.
- **c.** kingdom, domain, and order.
- **d.** kingdom, domain, and family.

______ **10.** The scientific name for the common house cat is *Felis domesticus*. What is its genus?

a. *Felis*
b. house cat
c. feline
d. *domesticus*

______ **11.** Today, what do scientists do when they encounter an organism that fits none of the four kingdoms?

a. destroy the organism
b. change the organism
c. create a new category
d. leave the new organism alone

12. Explain why Carolus Linnaeus simplified the naming of living things by giving each species a two-part scientific name.

____

____

____

13. What traits do scientists use to divide prokaryotes into groups?

____

____

____

14. Why do scientists classify a sponge as a simple animal and not a plant?

____

____

____

CRITICAL THINKING

15. Hundreds of years ago, scientists did not have microscopes and were unable to learn about microscopic organisms. Describe how this might have affected daily life back then.

____

____

____

____

____

____

16. Imagine you have traveled forward in time 2 million years to discover that Earth still supports life. Briefly describe the conditions on Earth. Describe an organism that is unchanged, one that has changed slightly, and a brand-new organism on Earth.

____

____

____

____

____

____

INTERPRETING GRAPHICS

Use the diagram below to answer questions 17 and 18.

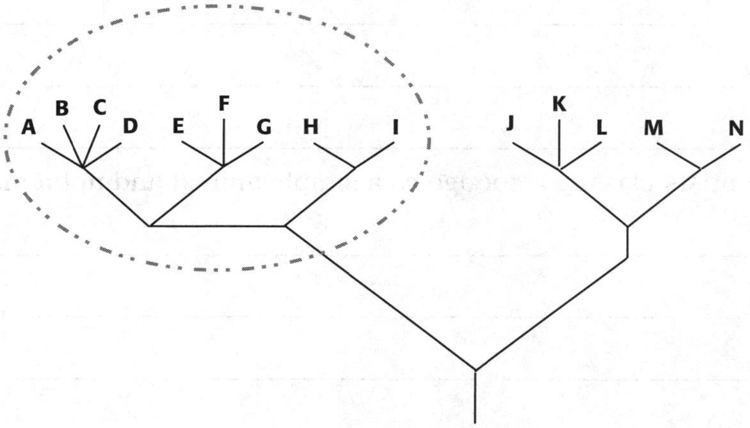

17. If species *A* through *I* are all in the same genus, what is similar about their scientific names?

____

____

____

18. Which organisms are more closely related: *I* and *J*, or *I* and *A*? Explain your answer.

____

____

____

CONCEPT MAPPING

19. Use the following terms to complete the concept map below:

domains	kingdoms
Protista	Eukarya

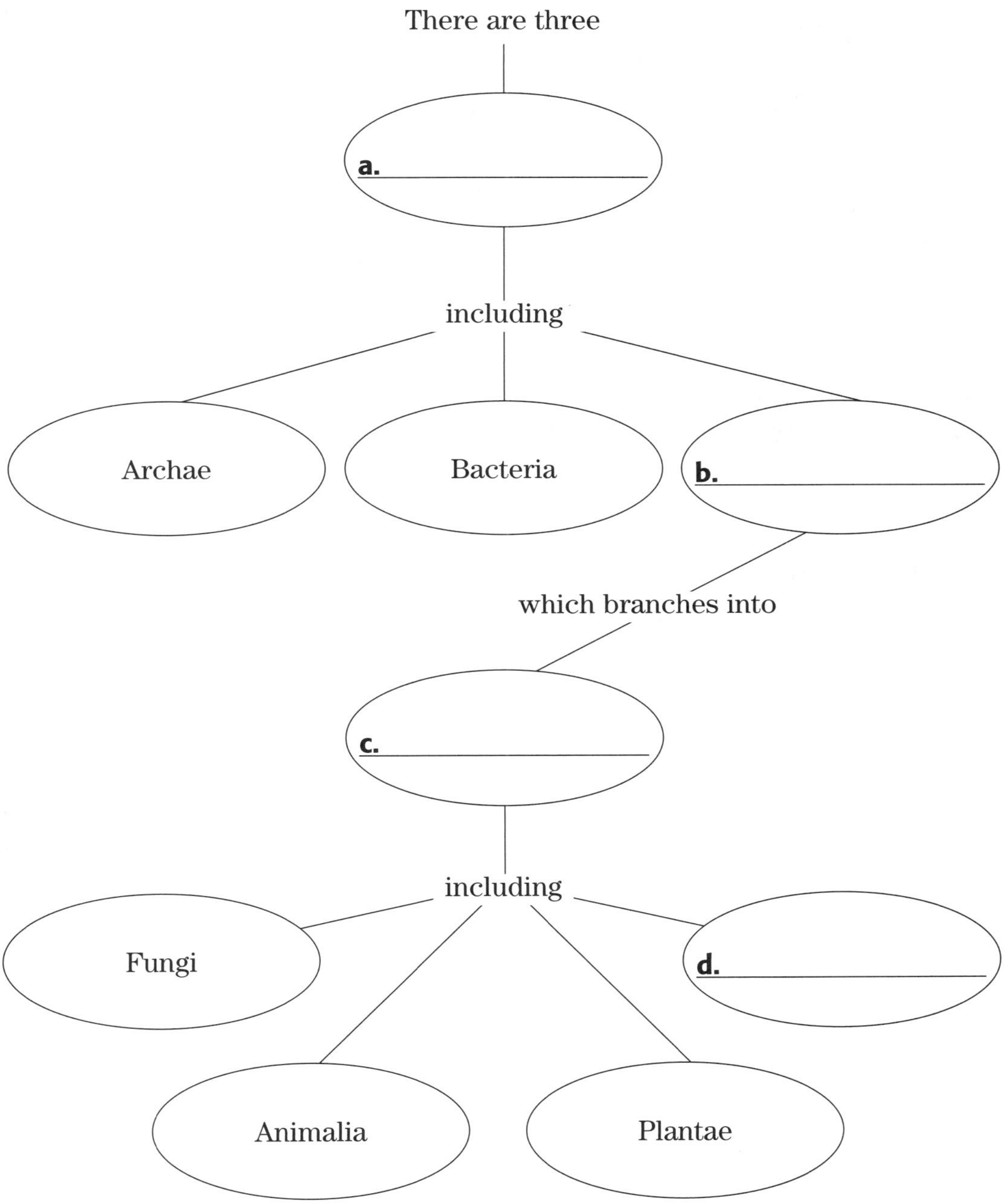

Performance-Based Assessment

Teacher Notes

PURPOSE

Students will create and use their own classification system to divide buttons into groups.

Jennifer Ford
Northridge Middle School
North Richland Hills, Texas

TIME REQUIRED

One 45-minute class period. Students will need 30 minutes to create the classification system and 15 minutes to answer the analysis questions.

RATING

Easy ← 1 2 3 4 → Hard

Teacher Prep–1
Student Set-Up–2
Concept Level–2
Clean Up–1

ADVANCE PREPARATION

You may wish to have students bring in buttons from home. Equip each student activity station with at least 20 buttons of different sizes, colors, numbers of holes, etc.

TEACHING STRATEGIES

This activity works best in groups of 3–4 students. Remind students that scientists classify organisms according to the characteristics they share.

Evaluation Strategies

Use the following rubric to help evaluate student performance.

Rubric for Assessment

Possible points	Appropriate use of materials and equipment (20 points possible)
20–14	Successfully completes activity; safe and careful handling of materials and equipment
13–7	Task is generally complete; appropriate use of materials and equipment
6–1	Attempts to complete activity yield inadequate results; apparent lack of skill
	Execution of activity (40 points possible)
40–30	Excellent classification system with superior detail; clear understanding of concepts and instructions; clear and precise descriptions of categories
29–20	Adequate classification system with moderate level of detail; clear understanding of concepts and instructions; mostly clear descriptions of categories
19–10	Mostly complete classification system with little detail; some understanding of concepts
9–1	Incomplete classification system with no detail; little understanding of concepts and instructions; no descriptions of categories
	Quality and clarity of observations (40 points possible)
40–27	Clear and detailed responses that show superior knowledge of classification systems; correct terminology and examples in responses
26–13	Adequate responses that show good knowledge of classification systems; some correct terminology; minor difficulty in responses
12–1	Incomplete or poor responses that show little knowledge of classification systems; incorrect terminology; unclear responses

Name ______________________ Class ______________ Date ______________

Assessment

SKILLS PRACTICE

Performance-Based Assessment

OBJECTIVE

You've read about how scientists classify living things. Now you will have a chance to create and use your own classification system to divide buttons into groups. Scientists had to make important choices when they designed the classification system used for living things that is in place today. You will make similar choices in this activity.

KNOW THE SCORE!

As you work through the activity, keep in mind that you will be earning a grade for the following:

- how well you work with the materials and equipment (20%)
- how accurately you follow the instructions to create your own system of classification (40%)
- how well you complete the analysis questions (40%)

Using Scientific Methods

ASK A QUESTION

How can I classify buttons into classes and families?

MATERIALS AND EQUIPMENT

- buttons, large selection
- containers, such as jars, boxes, or box lids
- labels

SAFETY INFORMATION

- Do not eat or drink anything in the laboratory.

FORM A HYPOTHESIS

1. How can a collection of buttons be classified into groups?

____

____

____

____

TEST THE HYPOTHESIS

2. Spread the buttons out so all can be seen at once.

3. Choose some interesting features by which you can divide the buttons into "classes." Separate the buttons, and give each class a descriptive name.

4. Label each container with the class of button it will hold, and put the buttons in the correct containers.

5. Now that you have your buttons divided into classes, decide how you can divide your buttons further, into families. Separate the buttons, and give each family a descriptive name.

ANALYZE THE RESULTS

1. How many classes of buttons do you have?

____

2. List the classes here, and describe the major characteristic of each class.

____

____

____

3. Were there any buttons that ended up in a class by themselves? Why?

____

____

4. How many families of buttons do you have?

____

5. List the families here, and describe the major characteristics of each.

____

____

____

____

DRAW CONCLUSIONS

6. If you were asked to divide the families further, what would this next level of classification be called?

____

7. What characteristics might you use to define this next division?

____

____

____

BIG IDEA QUESTION

8. How did the process of classifying buttons parallel the process scientists follow to classify organisms?

____

____

____

____

____

____

Assessment

Standards Assessment

Teacher Notes and Answer Key

To provide practice under more realistic testing conditions, give students 20 min to answer all of the questions in this assessment.

QUESTION NUMBER	CORRECT ANSWER	STANDARD
1	*B*	7.1 (Mastering)
2	*D*	7.3.d (Supporting)
3	*B*	7.3.d (Supporting)
4	*C*	7.3.d (Supporting)
5	*B*	7.3.d (Supporting)
6	*C*	7.3.d (Mastering)
7	*B*	7.3.d (Mastering)
8	*D*	7.3.d (Mastering)
9	*A*	7.3.d (Mastering)
10	*C*	7.3.d (Mastering)
11	*C*	7.1.a (Mastering)
12	*B*	7.3.e (Exceeding)
13	*B*	7.2.e (Mastering)
14	*D*	6.5.a (Supporting)
15	*A*	7.3.b (Mastering)

TEST DOCTOR

The following Standards Assessment questions have been diagnosed by the Test Doctor. Find out what might be causing your students' "ailing" answers. Each Test Doctor is followed by a diagnostic teaching tip to help you address students' learning needs.

Question 1 *asks students to define a word in context.*

A Incorrect A microscope does not hide cells.

B Correct. *Observable* and *visible* are interchangeable in most sentences. They both refer to things that can be seen.

C Incorrect. *Separable* means "able to be separated"; a microscope does not separate things.

D Incorrect. The details of a cell are not driven by a microscope.

Diagnostic Teaching Tip: Students who had difficulty answering this question correctly might understand the meaning of visible better by relating it to personal experience. Ask them to list things that are visible, such as celestial bodies in the sky before dark and after dark.

Question 2 *asks students to match the given definition with a word.*

A Incorrect. *Adapt* means "to change something to make it more suitable to a new situation or environment."

B Incorrect. *Interpret means* "to explain the meaning of something."

C Incorrect. *Display* means "to show something."

D Correct. *Construct* means "to build something by putting parts together."

Diagnostic Teaching Tip: Students who have difficulty answering this question might benefit from playing a game in which they match definitions with words. Have each student look up two words and write them on index cards. Have them write the definitions of the words on separate index cards. Put students in small groups to play matching games with their cards.

Question 3 *asks students to indicate the meaning of a word in context.*

A Incorrect. This sentence does not use *function* in a mathematical context.
B Correct: The branching diagram is a specialized tool that is used to show how different kinds of organisms are related to each other.
C Incorrect. *Function* does not mean "a characteristic that depends on another."
D Incorrect. The definition of a function as a social gathering does not make sense in the context of using a branching diagram to classify organisms.

Diagnostic Teaching Tip: Students who struggle with this question might benefit from practicing understanding words with multiple definitions. For example, have students find two possible meanings of the following words: *wind, structure, organ, tissues, construct,* and *compose.* Have them list one other word that has multiple meanings.

Question 4 *asks students to choose the correct collocation of words in context.*

A Incorrect. The preposition *into* is not used with the verb derive.
B Incorrect. The preposition *with* is not used with *derive.*
C Correct. The scientific name *Tyrannosaurus rex* is derived from two Greek words.
D Incorrect. *Derive* is not used with the preposition *on.*

Diagnostic Teaching Tip: Students who struggle to answer this question might benefit from practice using *derived from* in other contexts. Have students brainstorm a list of things that are derived from other things. Have them write a word equation using arrows to show which substance, object, or organism is derived from what.

Question 5 *asks students to recognize the levels in a classification system.*

A Incorrect. These categories are not levels in the formal scientific classification system used today.
B Correct. Families, genera, and species are levels of the formal scientific classification system used today.
C Incorrect. Endangered organisms and organisms that are not endangered are found at all levels of classification; these categories are not levels in the formal scientific classification systems used today.
D Incorrect. These are kinds of habitats, not levels in a scientific classification system.

Diagnostic Teaching Tip: Students who have difficulty answering this question correctly might benefit from making a concept map. Have them use specific organisms to show the relationships among the levels of classification.

Question 6 *asks students to interpret a branching diagram.*

A Incorrect. The lack of retractable claws is what separates the brown bear from the right side of the branching diagram.

B Incorrect. The house cat has retractable claws, but it is not the only such animal in this diagram.

C Correct. The diagram shows that the lion and the house cat both have retractable claws.

D Incorrect. The platypus and the brown bear are the two animals in the diagram that do not have retractable claws.

Diagnostic Teaching Tip: Students who have difficulty answering this question correctly might benefit from a group discussion about how to interpret a branching diagram. Provide two examples of branching diagrams, one containing plants and the other containing animals. Have students suggest questions the diagrams can help them answer. Students who choose answer A or answer D may be uncertain which side of the branching diagram a given characteristic applies to.

Question 7 *asks students to demonstrate an understanding of branching diagrams.*

A Incorrect. An italicized name does not indicate where the fossil should be placed on the branching diagram.

B Correct. An immediate ancestor of the brown bear should be placed on the same branch as the brown bear.

C Incorrect. The fossil on a different branch than the brown bear indicates that the fossil was not an immediate ancestor of the brown bear.

D Incorrect. An organism appearing at the base of a branching diagram indicates that is it a distant ancestor of the brown bear, not an immediate ancestor.

Diagnostic Tip: Students who struggle to answer this type of question might be confused about how a branching diagram is constructed. Give the student a list of animals. Have them list the characteristics of each animal, and then construct a branching diagram based on those characteristics.

Question 8 *asks students to match a famous scientist with his contribution.*

A Incorrect. The Chicabo people know of 360 species of plants living in Peru.

B Incorrect. Oliver Zompro is famous for discovering an insect that had not been previously classified.

C Incorrect. Michael Fay spent fifteen years observing and classifying animals in Africa.

D Correct. Carolus Linnaeus developed the binomial nomenclature system for classifying organisms. The system scientists use today is based on Linnaeus's system.

Diagnostic Teaching Tip: Students who have difficulty answering this question might benefit from finding out more about one of the groups or people mentioned above. Ask the students to write a short report about one of the answer selections and include a bibliography to cite the sources used.

Question 9 *asks students to demonstrate understanding of the scientific classification of organisms.*

A Correct. A single-celled organism without a nucleus is a prokaryote, so it must fall into one of the two prokaryotic domains, Archaea or Bacteria. Archaea can survive in extreme environments such as hot springs, so this fossil is most likely a member of Archaea.

B Incorrect. Although plant cells do have cell walls, the absence of a nucleus means this fossil is a prokaryote and therefore is not a member of group Plantae because Plantae is a kingdom in the domain Eukarya.

C Incorrect. Protista is a kingdom in the domain Eukarya. All protist cells have nuclei. This fossil does not have a nucleus and therefore cannot be a protist.

D Incorrect. Fungi is a kingdom in the domain Eukarya. All fungal cells have nuclei. This fossil does not have a nucleus and therefore cannot be a fungus.

Diagnostic Teaching Tip: Students who have difficulty answering this question correctly might benefit from practice classifying organisms using the eight-level system. Have students select three organisms and list their classifications in order from general to specific.

Question 10 *asks students to interpret a branching diagram.*

A Incorrect. Hibiscus is the organism shown farthest to the right on the diagram, so it is one of two kinds of organisms that has evolved most recently.

B Incorrect. Fern is not shown on the leftmost branch of the diagram, so it is not the oldest organism in the diagram.

C Correct. Moss is listed on the leftmost branch of the organism, so it is the oldest organism in the diagram.

D Incorrect. According to the branching diagram, the pine tree is as recently evolved as the hibiscus.

Diagnostic Teaching Tip: Students who struggle to answer this question might benefit from creating their own branching diagrams. Brainstorm people, places, things, and ideas that can be broken into groups according to similar characteristics, such as books in a library, departments in a store, or students in a school. Have students choose one topic to diagram.

Question 11 *asks students to demonstrate an understanding that cells function similarly in all organisms.*

A Incorrect. All living things are made up of cells.
B Incorrect. Animal cells cannot make their own food.
C Correct. Cells function similarly in all organisms. For example, all cells make proteins.
D Incorrect. Some single-celled organisms do not have nuclei. Organisms without nuclei are called prokaryotes.

Diagnostic Teaching Tip: Students who struggle with this item might benefit from reviewing the similarities between prokaryotes and eukaryotes. Instruct students to create a diagram of a prokaryotic cell and a eukaryotic cell. On their diagrams, have them list several examples of organisms that have prokaryotic cells and organisms that have eukaryotic cells. All plants and animals have eukaryotic cells.

Question 12 *asks students about causes of evolution and extinction.*

A Incorrect. Hurricanes are likely to cause the death of individual plants and animals but not the extinction of large numbers of species.
B Correct. Drastic climate change, such as the change that occurs at the onset of an ice age, can cause of the extinction of large numbers of species.
C Incorrect. Floods are not likely to cause many species of organisms to become extinct.
D Incorrect. Wars cause the death of humans but do not routinely cause the widespread extinction of species.

Diagnostic Teaching Tip: Students who struggle to answer this question might benefit from doing research at the library and on the Internet about events that have caused drastic climate change. Ask them to share what they learn about the effects of one of these events on a particular species of plant or animal.

Question 13 *asks students to recognize the difference between where DNA is found in prokaryotic and eukaryotic cells.*

A Incorrect. DNA forms a loose loop within the cells in prokaryotes, which lack nuclei. Algae, however, are eukaryotic.
B Correct. Algal DNA, like that of all eukaryotes, is found in the nucleus in long strands called chromatin.
C Incorrect. DNA is found inside the nucleus of eukaryotic cells.
D Incorrect. The endoplasmic reticulum does not contain DNA.

Diagnostic Teaching Tip: Students who struggle to answer this question correctly might benefit from reviewing the differences between prokaryotes and eukaryotes, as well as how algae is classified as a eukaryote.

Question 14 *asks students to demonstrate an understanding of the relationship between different types of organisms and the flow of energy in the biosphere.*

A Incorrect. Once energy from sunlight has been converted to chemical energy, it cannot reenter the biosphere as sunlight. The sun is the only source of sunlight.

B Incorrect. Sunlight cannot be used directly as an energy source by consumers. As the word "consumers" implies, consumers must consume other organisms to obtain the energy that they need.

C Incorrect. Animals are consumers and cannot use the sun's energy directly. Animals get their energy by consuming other organisms.

D Correct. Energy enters the biosphere as sunlight. Plants convert the energy into chemical energy. The energy passes to animals that eat the plants, and eventually to the other organisms that help break down dead organisms.

Diagnostic Teaching Tip: Students who struggle to answer this question might benefit from examining a few different organisms and arranging them according to their places in the flow of energy through the biosphere. Keep in mind that some plants, such as the Venus flytrap, may qualify as both producers and consumers, and some animals, such as insects, may qualify as both consumers and decomposers. Ask students why they think our formal system of classification is not based on organisms' roles in the energy cycle.

Question 15 *asks students to demonstrate an understanding of Charles Darwin's theory of natural selection.*

A Correct. Darwin's conclusion is called natural selection. Environmental and ecological conditions "select" certain heritable characteristics of plants and animals for survival and reproduction.

B Incorrect. Extinction occurs when all the members of a species die out.

C Incorrect. Genetic homogeneity means that members of a species are alike on a genetic level.

D Incorrect. Darwin's idea takes ecological factors into account, but they are only part of his theory.

Diagnostic Teaching Tip: Students who struggle with this question might benefit from making a concept map of each of the four terms given in the question. Then have students erase the term from the center of the map and allow other students to guess the term that belongs in the center of the concept map.

Name ______________________ Class ______________ Date ______________

Assessment

Standards Assessment

REVIEWING ACADEMIC VOCABULARY

______ **1.** In the sentence "The nucleus of a human cell is visible only with a microscope," what does the word *visible* mean?
A hidden
B observable
C separable
D driven

______ **2.** Which of the following words means "to build something by putting parts together"?
A adapt
B interpret
C display
D construct

______ **3.** In the sentence "The function of a branching diagram is to show how different kinds of organisms are related to each other," what does the word *function* mean?
A a relationship between mathematical sets
B a specialized activity of a tool or system
C a characteristic that depends upon another
D a social gathering or formal ceremony

______ **4.** Which of the following sets of words best completes the following sentence: "The scientific name *Tyrannosaurus rex* _____ two Greek words that mean 'tyrant lizard'"?
A derives into
B is derived with
C is derived from
D derives on

REVIEWING CONCEPTS

______ **5.** Classification is a tool that is used to study living and extinct organisms. Which of the following are levels in this system?
A living and nonliving
B families, genera, and species
C endangered and not endangered
D land, marine, and freshwater organisms

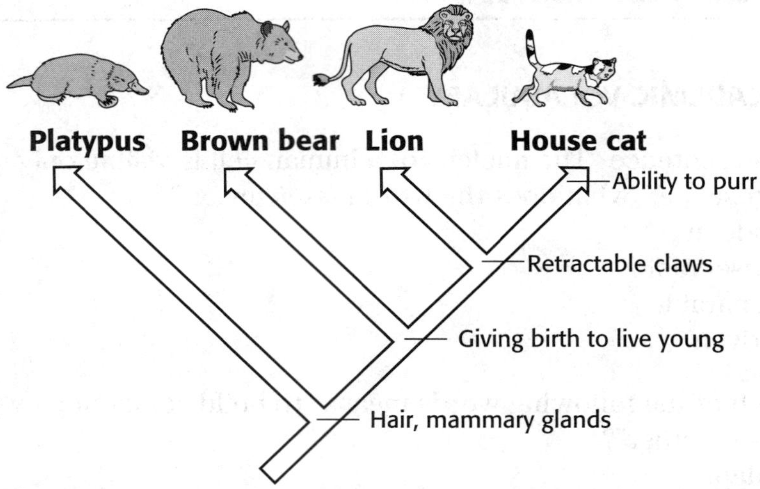

______ **6.** The branching diagram above shows similarities and differences between four mammals. Which of the following mammals has retractable claws?
A only the brown bear
B only the house cat
C the lion and the house cat
D the platypus and the brown bear

______ **7.** A scientist discovers a fossil of a new mammal that was an immediate ancestor of the brown bear. How would this fossil be shown in a branching diagram?
A Its name would be italicized.
B Its name would be placed on a branch leading to the brown bear.
C Its name would be placed at the tip of a new branch next to the brown bear.
D Its name would appear at the base of a branching diagram showing all bears.

______ **8.** Scientists today use an eight-level classification system. Who developed the original system on which the current classification system is based?
A the Chicabo people
B Oliver Zompro
C Michael Fay
D Carolus Linnaeus

______ **9.** The fossil of an organism is found near a hot spring. This single-celled organism has a cell wall and no nucleus. Into which of the following groups would you classify this organism?
A domain Archaea
B kingdom Plantae
C kingdom Protista
D kingdom Fungi

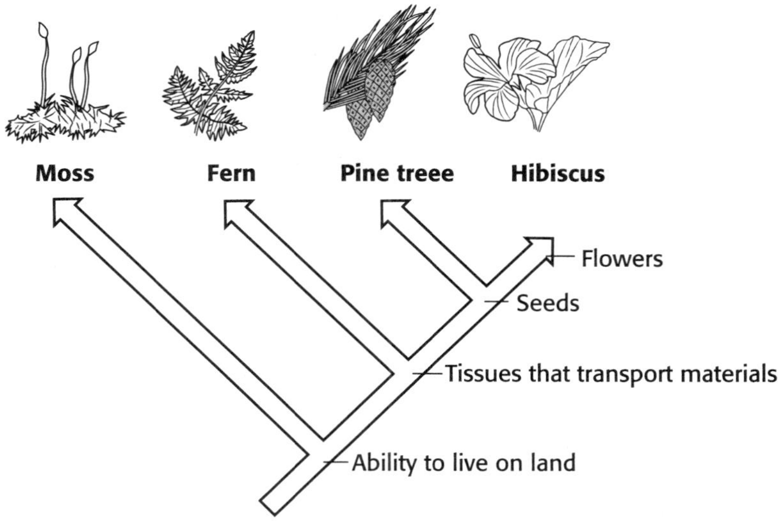

______ **10.** The branching diagram above shows the relationships between four groups of plants. According to the diagram, which group of plants is the oldest?
A hibiscus
B fern
C moss
D pine tree

REVIEWING PRIOR LEARNING

______ **11.** Which of the following statements about cells is true?
A Some organisms have no cells.
B All cells make their own food.
C All cells function similarly.
D All cells have a nucleus.

______ **12.** Which of the following events is most likely to lead to the extinction of the largest number of species?
A a hurricane
B an ice age
C a flood
D a war

______ **13.** An algal cell's DNA is found
A in a loose loop within the cell.
B in long strands called *chromatin.*
C in the cytoplasm outside the nucleus.
D in the cell's endoplasmic reticulum.

______ **14.** Energy flows through the biosphere

A from decomposers to producers to consumers to sunlight.
B from sunlight to animals to producers.
C from sunlight through animals to plants.
D from sunlight through producers to consumers.

______ **15.** Darwin developed the following theory: Organisms that have heritable characteristics that are best suited to the organisms' environment survive and pass those characteristics on to their offspring more successfully than organisms that are not well suited to the environment do. What is this theory called?

A natural selection
B natural extinction
C genetic homogeneity
D uniformitarianism

Analyzing a Branching Diagram

Teacher Notes

In this activity, students will learn what a branching diagram represents (covers standards 7.3.d and 7.7.d). You may want to give students who are struggling with the activity hints on how the branching diagram incorporates variation of the subject along a timeline.

Explore Activity

Analyzing a Branching Diagram

Scientists use branching diagrams to show how different kinds of organisms are related to each other. The branching diagram below shows one way that clothes can be organized into groups.

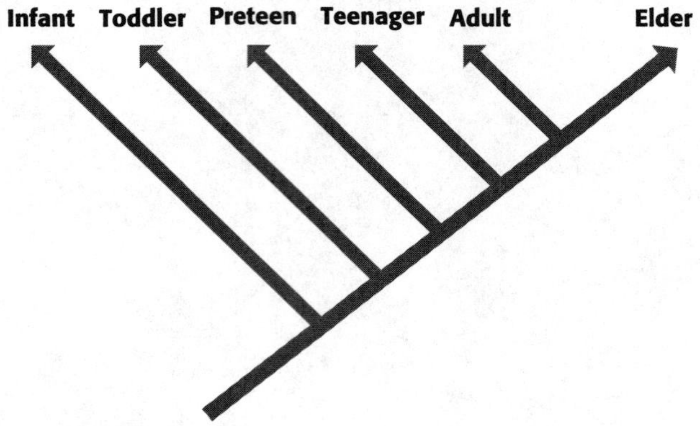

PROCEDURE

1. Redraw the diagram in the space below.
- Leave extra space between each branch.

2. On the branch that slopes towards the right, draw a mark between each of the other branches in your diagram.
- Label each mark with something that changed between the branches that are on either side of the mark.

ANALYSIS

3. Complete the following statement:
The diagram shows how clothing can be divided into groups based on

____

4. Which branch of clothing category do you fit in? (Hint: What age group are you in?)

____

5. Which categories of clothing have you already worn? (Hint: What age groups have you passed?)

____

6. Choose one group of clothing in the diagram.

- How could this group be split into two new groups?

____

____

- Show these two new groups on your diagram.

Explore Activity **DATASHEET B**

Analyzing a Branching Diagram

Scientists use branching diagrams to show how different kinds of organisms are related to each other. The branching diagram below shows one way that clothes can be organized into groups.

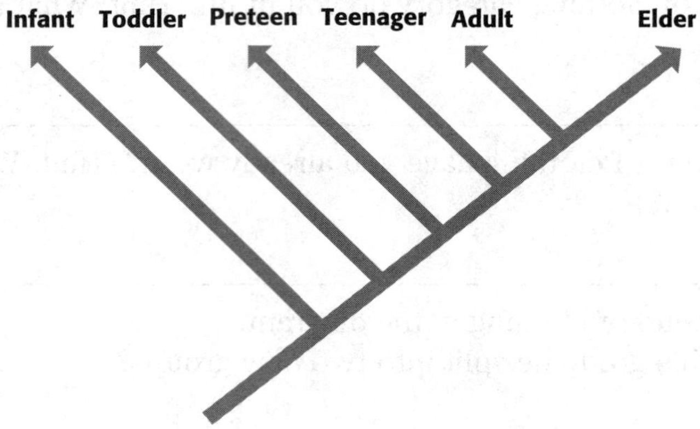

PROCEDURE

1. Redraw the diagram.

2. On the branch that slopes towards the right, draw a mark between each of the other branches in your diagram. Label each mark with what changed between the branches that are on either side of the mark.

ANALYSIS

3. Write a statement that explains what you think the diagram shows.

____

____

4. Which branch of clothing category do you fit in?

____

5. Which categories of clothing have you already worn?

____

6. Can you think of how one of these groups of clothing can be split into two new groups? Show the two new groups on your diagram.

____

Name ______________________ Class ______________ Date ____________

Explore Activity

DATASHEET C

Analyzing a Branching Diagram

Scientists use branching diagrams to show how different kinds of organisms are related to each other. The branching diagram below shows one way that clothes can be organized into groups.

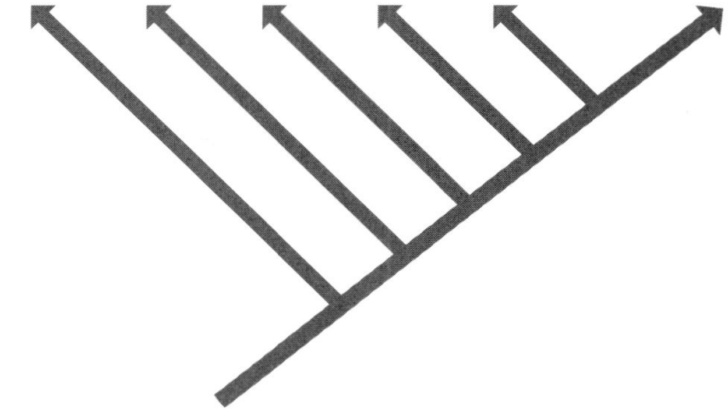

PROCEDURE

1. Redraw the diagram in the space below.

2. On the branch that slopes towards the right, draw a mark between each of the other branches in your diagram. Label each mark with what changed between the branches that are on either side of the mark.

ANALYSIS

3. Explain what you think the diagram shows.

____

____

____

____

4. Which branch of clothing category do you fit in?

____

____

5. Which categories of clothing have you already worn?

____

____

6. Can you think of how one of these groups of clothing can be split into two new groups? Show the two new groups on your diagram.

____

____

Constructing a Branching Diagram

Teacher Notes

In this activity, students draw their own branching diagram (covers standard 7.3.d).

Name ______________________ Class ______________ Date ____________

Quick Lab

DATASHEET A

Constructing a Branching Diagram

PROCEDURE

1. Draw a diagram similar to the one in **Figure 2** in your book in the space below. On the diagram, write "frog", "snake", "kangaroo", and "rabbit" in this order on the branches, from left to right.

2. Draw a mark on the line right before the frog branches off.
- Think of one major change that happened before the frog evolved.
- All the animals should share this characteristic.
- Write the characteristic next to the mark.

____

3. Make a mark showing major changes between each type of organism.
- Make a mark on the line between the frog and snake branches.
- Think of a major change that happened between the frog and the snake.
- The snake, kangaroo, and rabbit should share this characteristic.
- Write the characteristic next to the mark.

____

- Make a mark on the line between the snake and kangaroo branches.
- Think of a major change that happened between the snake and the kangaroo.
- The kangaroo and rabbit should share this characteristic.
- Write the characteristic next to the mark.

____

- Make a mark on the line between the kangaroo and rabbit branches.
- Think of a major change that happened between the kangaroo and the rabbit.
- The rabbit should have this characteristic.
- Write the characteristic next to the mark.

____

4. The characteristics listed by the marks are shared derived characteristics of all the animals that branch off above that mark. The more derived characteristics organisms share, the more closely related they are.
- Are rabbits more closely related to frogs and snakes, or to kangaroos? How do you know?

____

____

____

____

Quick Lab **DATASHEET B**

Constructing a Branching Diagram

PROCEDURE

1. Construct a diagram similar to the one in **Figure 2** in your book. On the diagram, write “frog”, “snake”, “kangaroo”, and “rabbit” in this order, from left to right.

2. Think of one major change that happened before the frog evolved.
3. For the last three organisms, write a change that happened between one of these organisms and the other two in your diagram.
4. How does this diagram show that the organisms are related?

____

____

____

____

Name ______________________ Class ______________ Date ____________

Quick Lab

DATASHEET C

Constructing a Branching Diagram

PROCEDURE

1. In the space below, construct a diagram similar to the one in **Figure 2** in your book. On the diagram, write “frog”, “snake”, “kangaroo”, and “rabbit” in this order, from left to right.

2. Think of one major change that happened before the frog evolved, and mark it in the appropriate place on your diagram.
3. Think of a major change that happened between each of the organisms, and write the changes in the appropriate places on your diagram.
4. Explain how this branching diagram shows how the organisms are related.

____

____

____

____

Fossils and Branching Diagrams

Teacher Notes

In this activity, students learn how to include a fossil into a branching diagram they have created. Students should classify the fossil based on their observation of the fossil's characteristics, such as the feathers and a lizard-shaped or dinosaur-shaped head (covers standard 7.3.d).

Name ______________________ Class ______________ Date ____________

Quick Lab

DATASHEET A

Fossils and Branching Diagrams

Try this activity to learn how you can study a fossil and include it in a branching diagram with modern organisms.

TRY IT!

1. The image in your book is a photo of a fossil of an organism called an archaeopteryx.

2. What are some characteristics of the archaeopteryx?

- Does the archaeopteryx appear to have a backbone?

____

- Does the archaeopteryx appear to have feathers or scales?

____

- Does the archaeopteryx appear to have a head shaped like a bird, or like a dinosaur?

____

3. Draw a branching diagram of the animal kingdom.

- Include fish, amphibians, reptiles, birds, and apes in the correct order.
- Among these organisms, fish form the oldest group.
- Amphibians came next and were followed by reptiles and then birds.
- Among these organisms, apes are the youngest group.

4. Consider the characteristics of each group of organisms.
- Fish have an endoskeleton and scales, and they live in water.
- Amphibians have an endoskeleton, and adults live on land, but their eggs need water to survive.
- Reptiles have an endoskeleton and scales, and they live on land.
- Birds have an endoskeleton and feathers, and they live on land.
- Apes have an endoskeleton and fur, and they live on land.
- Compare the characteristics of the archaeopteryx to the characteristics of the other animal groups.
- Decide where the archaeopteryx would fit in your branching diagram.
- Add its branch to your diagram. Make sure the branch does not extend to the tip of the other branches.

THINK ABOUT IT!

5. Look at your branching diagram. How can you tell that the archaeopteryx is an extinct organism? (Hint: Look at how long the archaeopteryx branch is compared to organisms that still exist today.)

____

____

6. Look at your branching diagram.
- Which kinds of organisms are younger (evolved more recently) than the archaeopteryx?

____

____

____

- Which kinds of organisms are older than the archaeopteryx?

____

____

____

Name ______________________ Class ______________ Date ______________

Quick Lab

DATASHEET B

Fossils and Branching Diagrams

Try this activity to learn how you can study a fossil and include it in a branching diagram with modern organisms.

TRY IT!

1. The image in your textbook is a photo of a fossil of an organism called an archaeopteryx.
2. Create a simple list of characteristics that describe the archaeopteryx.

____

____

3. On a separate sheet of paper, construct a branching diagram of the animal kingdom. Include fish, amphibians, reptiles, birds, and apes in the branching diagram. (Hint: Among these organisms, fish evolved earliest. Amphibians formed the next group, and they were followed by reptiles and then birds. Among these organisms, apes are the most recent group.)
4. Consider the characteristics of each group of organisms. Decide where the archaeopteryx would fit in your branching diagram, and add it to your diagram.

THINK ABOUT IT!

5. Using your branching diagram, how can you tell that the archaeopteryx is extinct?

____

____

____

____

6. Which kinds of organisms are more recent than the archaeopteryx? Which kinds of organisms evolved before the archaeopteryx?

____

____

____

____

____

Name ______________________ Class ______________ Date ______________

Quick Lab

DATASHEET C

Fossils and Branching Diagrams

Try this activity to learn how you can study a fossil and include it in a branching diagram with modern organisms.

TRY IT!

1. The image in your book is a photo of a fossil of an organism called an archaeopteryx.
2. Create a simple list of characteristics that describe the archaeopteryx.
3. On a separate sheet of paper, construct a branching diagram of the animal kingdom. Include fish, amphibians, reptiles, birds, and apes in the branching diagram.
4. Consider the characteristics of each group of organisms. Decide where the archaeopteryx would fit in your branching diagram and add it to your diagram.

THINK ABOUT IT!

5. How does your branching diagram show that the archaeopteryx is an extinct organism?

____

____

____

6. Which kinds of organisms are more recent than the archaeopteryx? Which kinds of organisms evolved before the archaeopteryx? How do you know?

____

____

____

____

____

____

____

____

Skills Practice Lab

DATASHEET

Grouping Life-Forms by Their Characteristics

Teacher Notes

In this activity, students use the shared derived characteristics of several life-forms to create their own branching diagram. The activity also requires students to include the fossil of a life-form in their branching diagram (covers standards 7.3.d and 7.7.d).

CLASSROOM TESTED & APPROVED

Sushma Kashyap
Alvarado Intermediate School
Rowland Heights, California

TIME REQUIRED

One 45-minute class period

LAB RATINGS

Easy ← 1 2 3 4 → Hard

Teacher Prep–1
Student Set-Up–1
Concept Level–3
Clean Up–1

PREPARATION NOTES

Some students may have greater success at this lab if they are given photocopied cut outs of the life-forms that they can manipulate. Be sure that students understand that changes in the complexity of groups of organisms are likely to be linear. Thus, in deciding which life-forms came first and which life-forms came later, students may logically assume that life-forms evolved from having fewer body segments to having more body segments.

Name ______________________ Class ______________ Date ____________

Skills Practice Lab

DATASHEET A

Grouping Life-Forms by Their Characteristics

You are a crew member and a biologist on the USS *Adventure*. The USS *Adventure* has been on a 5-year mission to collect life-forms and fossils from outside the solar system. On the voyage back to Earth, your ship went through a meteor shower. The meteor shower damaged several of the rooms containing the alien life-forms and the fossils you collected. Now, there are only two undamaged rooms in your starship for the life-forms. Assuming that life-forms with similar characteristics can live safely together, use observable characteristics to group the life-forms into these two rooms.

Life-form 1

Life-form 2

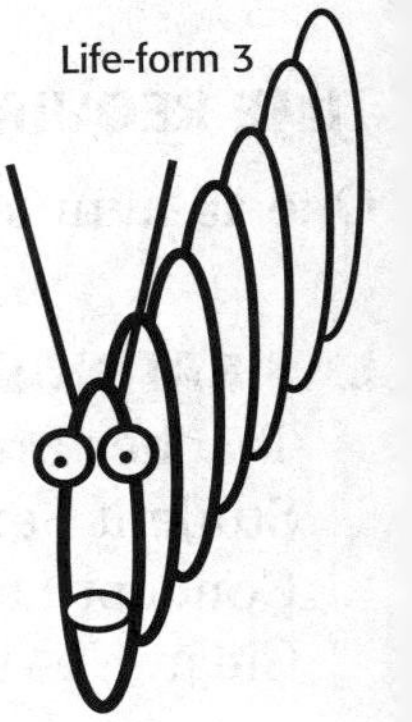
Life-form 3

Life-form 4

Life-form 5

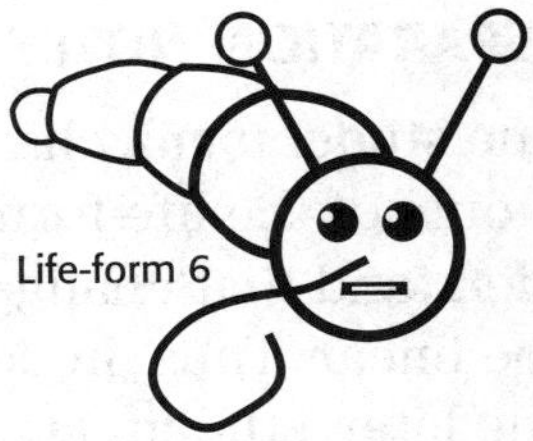
Life-form 6

Fossil of life-form 7

OBJECTIVES

Observe the characteristics of different life-forms.

Identify how life-forms can be grouped based on their characteristics.

Construct a branching diagram.

MATERIALS

- paper
- pencil

PROCEDURE

1. Use the following data table to complete steps 2, 3, and 6.

LIFE-FORM CHARACTERISTICS

Life-form	Number of body segments	Number of antennae	Room assignment
1			
2			
3			
4			
5			
6			
7			

2. Life-forms 1 through 7, and life-form CC9 are shown in your book. For life-forms 1–7, observe the shared derived characteristics.
- Shared derived characteristics are characteristics that two or more life-forms share with their most recent common ancestor.
- Shared derived characteristics observed should include the number of body segments and the number of antennae in each life-form.
- Fill in the columns about the shared-derived characteristics of the life-forms based on your observations. For example, life-form 2 has one body segment and one antenna.
- Use your observations about the number of body segments to divide the life-forms into two main groups: group 1 and group 2.
- Your room assignments should correspond to these groups.

3. Fill in the last column of your table with the room assignments for each life-form, placing members of group 1 into room 1 and members of group 2 into room 2.

4. On a separate piece of paper make a branching diagram based on the characteristics you have recorded in the table.
- Start with life-form 1 at the base of the branching diagram.
- Since you have life-forms in two compartments, you should have two branches splitting off from life-form 1.
- Include the fossil of life-form 7 in the branching diagram.

5. To decide which life-forms evolved before other related life-forms, use the information about the number of body segments or the number of antennae they have.
   - For example, life-forms with more body segments may have evolved more recently than life-forms with fewer body segments.

6. The USS *Adventure* makes one more stop before returning home. On planet X437 you discover another interesting life-form—the CC9, shown in your book.
   - Based on your previous grouping of life-forms, decide if you can include CC9 in one of the rooms for the trip to Earth.

CC9

ANALYZE THE RESULTS

7. **Identifying Patterns** Describe the life-forms in room 1. How are they similar?

   ____

   ____

   How are they different?

   ____

   ____

8. **Identifying Patterns** Describe the life-forms in room 2. How are they similar?

   ____

   ____

   How do they differ from the life-forms in room 1?

   ____

   ____

9. **Interpreting Information** Which life-forms is the fossil related to? (Hint: Use your branching diagram, and look at number of body segments.)

   ____

   ____

   ____

   ____

DRAW CONCLUSIONS

10. Applying Conclusions Life-form CC9 does not have repeating body segments. In which room should life-form CC9 be placed?

____

____

____

11. Making Inferences Some of the life-forms have eyes, mouths, and legs.

- Could these characteristics have been used to make a branching diagram of these life-forms?

____

____

- Would the branching diagram be different from yours if one of these other characteristics were used to make it?

____

____

- How might using different characteristics in making branching diagrams cause problems for scientists?

____

____

____

BIG IDEA QUESTION

12. Identifying Patterns What characteristics are used to group living and extinct organisms?

____

____

- How do branching diagrams show evolutionary relationships?

____

____

____

Name ______________________ Class ______________ Date ______________

Skills Practice Lab

DATASHEET B

Grouping Life-Forms by Their Characteristics

You are a crew member and a biologist on the USS *Adventure*. The USS *Adventure* has been on a 5-year mission to collect life-forms and fossils from outside the solar system. On the voyage back to Earth, your ship went through a meteor shower. The meteor shower damaged several of the rooms containing the alien life-forms and the fossils that you collected. Now, there are only two undamaged rooms in your starship for the life-forms. Assuming that life-forms with similar characteristics can live safely together, use observable characteristics to group the life-forms into these two rooms.

Life-form 1

Life-form 2

Life-form 3

Life-form 4

OBJECTIVES

Observe the characteristics of different life-forms.

Identify how life-forms can be grouped based on their characteristics.

Construct a branching diagram.

Life-form 5

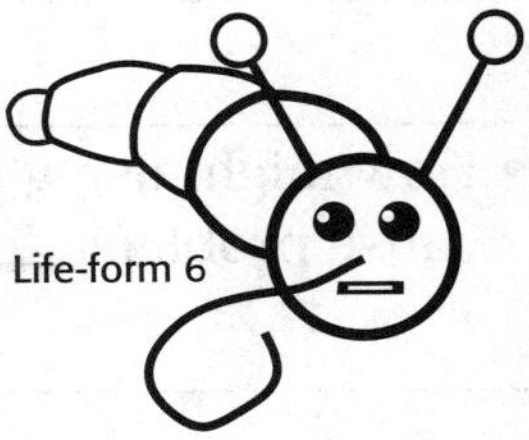
Life-form 6

Fossil of life-form 7

MATERIALS

- paper
- pencil

PROCEDURE

1. On a separate sheet of paper, make a data table similar to the one below for all of the life-forms shown.

LIFE-FORM CHARACTERISTICS

Life-form	Number of body segments	Number of antennae	Room assignment
1			
2			
3			

2. Fill in the columns about the shared derived characteristics of the life-forms based on your observations. (Remember that shared derived characteristics are characteristics that two or more life-forms share with their most recent common ancestor.) Use your observations to divide the life-forms and the remaining fossil into two main groups.

3. Fill in the last column of your table with the room assignments for each life-form.

4. On a separate sheet of paper, make a branching diagram based on the characteristics that you have recorded in the table. (Hint: Start with Life-form 1 at the base of the branching diagram. Because you have life-forms in two compartments, you are likely to have two branches splitting off from Life-form 1.) Also, include the fossil of Life-form 7 in the branching diagram.

5. In deciding the order in which life-forms evolved, use the information about the number of body segments or the number of antennae. For example, life-forms that have many body segments may have evolved more recently than life-forms that have fewer body segments.

6. The USS *Adventure* makes one more stop on the way home. On planet X437, you discover another interesting life-form—the CC9, which is shown at right. Based on your previous grouping of life-forms, decide if you can include CC9 in one of the rooms for the trip to Earth.

CC9

ANALYZE THE RESULTS

7. Identifying Patterns Describe the life-forms in room 1. How are they similar? How are they different?

____

____

____

____

____

____

8. Identifying Patterns Describe the life-forms in room 2. Are they similar to each other? How do they differ from the life-forms in room 1?

____

____

____

____

____

____

9. Interpreting Information To which lifeforms is the fossil related? (Hint: Use your branching diagram.)

____

____

____

DRAW CONCLUSIONS

10. Applying Conclusions In which room did you place life-form CC9? How did you decide?

____

____

____

____

11. Making Inferences What other characteristics could you use to make a branching diagram of these life-forms? Would the use of other information change the branching diagram? What difficulties can using other information cause for scientists?

____

____

____

____

____

BIG IDEA QUESTION

12. Identifying Patterns How do you group living and extinct organisms? How do branching diagrams show evolutionary relationships?

____

____

____

____

____

Name ______________________ Class ______________ Date ______________

Skills Practice Lab

DATASHEET C

Grouping Life-Forms by Their Characteristics

You are a crew member and a biologist on the USS *Adventure*. The USS *Adventure* has been on a 5-year mission to collect life-forms and fossils from outside the solar system. On the voyage back to Earth, your ship went through a meteor shower. The meteor shower damaged several of the rooms containing the alien life-forms and the fossils you collected. Now, there are only two undamaged rooms in your starship for the life-forms. Assuming that life-forms with similar characteristics can live safely together, use observable characteristics to group the life-forms into these two rooms.

Life-form 1

Life-form 2

Life-form 3

Life-form 4

Life-form 5

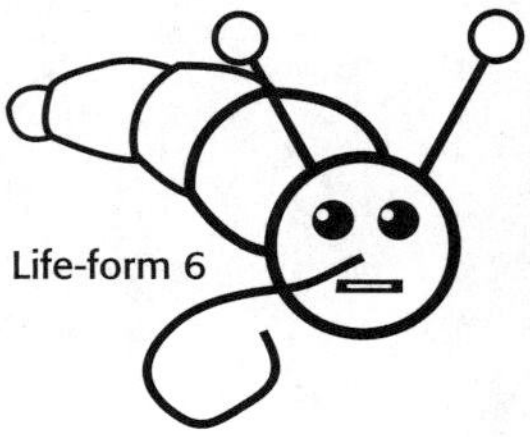
Life-form 6

Fossil of life-form 7

OBJECTIVES

Observe the characteristics of different life-forms.

Identify how life-forms can be grouped based on their characteristics.

Construct a branching diagram.

MATERIALS

For each group

- paper
- pencil

PROCEDURE

1. In the space below, make a data table similar to the one in your textbook for life-forms 1–7.

2. Fill in the columns about the shared derived characteristics of the life-forms based on your observations. Use your observations to divide the life-forms and fossil into two main groups.

3. Fill in the last column of your table with the room assignments for each life-form.

4. In the space below, make a branching diagram based on the characteristics you have recorded in the table. Also, include the fossil of life-form 7 in the branching diagram.

5. In deciding which life-forms evolved before other related life-forms, use the information about the number of body segments or the number of antennae they have. For example, life-forms with more body segments may have evolved more recently than life-forms with fewer body segments.

6. The USS *Adventure* makes one more stop before returning home. On planet X437, you discover another interesting life-form—the CC9, shown at right. Based on your previous grouping of life-forms, decide if you can include CC9 in one of the rooms for the trip to Earth.

CC9

ANALYZE THE RESULTS

7. Identifying Patterns Compare and contrast the life-forms in room 1.

____

____

____

8. Identifying Patterns Compare and contrast the life-forms in room 2. Contrast the life-forms in room 2 with the life-forms in room 1.

____

____

____

____

9. Interpreting Information To which life-form(s) is the fossil related?

____

____

____

DRAW CONCLUSIONS

10. Applying Conclusions In which room did you place life-form CC9? How did you decide?

____

____

____

____

11. Making Inferences What other characteristics could you use to make a branching diagram of these life-forms? Would the use of other information change the branching diagram? What difficulties can using other information cause for scientists?

____

____

____

____

____

____

BIG IDEA QUESTION

12. Identifying Patterns How do you group living and extinct organisms? How do branching diagrams show evolutionary relationships?

____

____

____

____

Communicating Through Written Reports

Teacher Notes

This activity asks students to use the appropriate writing style for scientific reports. To help students understand the differences in writing styles, use actual examples from a newspaper article, a scientific journal article, and an excerpt from a fictional story. Depending on the abilities of your students, you might ask some students to bring in their own research on the plesiosaur. You may wish to provide other students with photocopies of articles from the Internet, magazines, or science journals that they can use to write the report.

Name ______________________ Class ______________ Date ____________

Science Skills Activity

DATASHEET

Communicating Through Written Reports

INVESTIGATION AND EXPERIMENTATION

7.7.e Communicate the steps and results from an investigation in written reports and oral presentations.

TUTORIAL

After completing an investigation, scientists record and communicate the steps they followed as well as the results of their investigation. Communicating this information can be in the form of a written report. Use the following instructions to write a report based on the steps and results of an investigation.

1. **Gathering Information** First, gather the information that you collected during your investigation.
2. **Organizing Findings** Then, organize your information. Your report should have the following parts:
   - title
   - introduction
   - paragraph(s) on materials and methods used
   - discussion of your findings and conclusions
   - list of the sources of information used
3. **Communicating Findings**
   - In the introductory paragraph, describe the question that you want to answer.
   - In the materials and methods section, describe the steps you followed, the facts that you collected, and any observations that you made during the investigation.
   - In the discussion section, explain what you learned from your research. Make sure to state whether or not you found an answer to your question.
4. **Applying Concepts** The style of writing in scientific reports is different from the writing in opinion columns in newspapers or mystery stories. Writing in science must be *objective.* Objective writing is writing that includes only facts. Inferences are only included if there is enough evidence. Review your report to make sure that your writing is objective.

YOU TRY IT!

Procedure

Choose an extinct organism that interests you. Check with your teacher about your subject. Conduct research to find out about this organism. Write a report on your findings.

1. **Designing Questions** Think of a question that would help you learn about the fossils related to the organism you chose. For example, "Where and when did the Plesiosaur live?"

   ____

   ____

2. **Conducting Research** Use library resources and the Internet to find information about the organism you chose.

3. **Communicating Findings** Use the Tutorial to help you organize and communicate your findings on a separate piece of paper.

Analysis

4. **Applying Concepts** Is your report objective? Explain your position.

   ____

   ____

   ____

   ____

   ____

5. **Evaluating Sources** All of your conclusions were based on the information that you found in your research. What did you assume about the information that you found? What kinds of problems may arise from using sources that may not be accurate?

   ____

   ____

   ____

   ____

   ____

   ____

   ____

Answer Key

Directed Reading A

SECTION: SORTING IT ALL OUT

1. B
2. B
3. D
4. C
5. A
6. C
7. D
8. A
9. A
10. B
11. A
12. C
13. D
14. B
15. C
16. B
17. D
18. B
19. B
20. C
21. D
22. A

SECTION: DOMAINS AND KINGDOMS

1. A
2. D
3. B
4. B
5. B
6. A
7. C
8. D
9. C
10. C
11. A
12. B
13. E
14. D
15. C
16. C
17. A
18. B
19. A
20. D
21. C
22. B
23. A
24. C

Directed Reading B

SECTION: SORTING IT ALL OUT

1. Classification is the division of organisms into groups, or classes, based on specific characteristics.
2. B
3. B
4. B
5. C
6. A
7. D
8. A
9. Carolus Linnaeus
10. eight
11. They all have hair and mammary glands.
12. They give birth to live young.
13. A
14. D
15. B
16. phyla
17. genus
18. domain, kingdom, phylum, class, order, family, genus, species
19. scientific name
20. Before Linnaeus, scholars used names that had as many as 12 words to identify species. The system was difficult because scientists didn't always refer to organisms using the same names.
21. Carolus Linnaeus
22. genus
23. capital letter
24. lowercase letter
25. organism
26. Latin, Greek
27. specific
28. *T. rex*
29. The two organisms are likely closely related.
30. Miocene; *Megahippus*

SECTION: DOMAINS AND KINGDOMS

1. plants, animals
2. B
3. D
4. C
5. A
6. Protista
7. three
8. prokaryotes
9. bacteria
10. bacteria
11. archaea
12. archaea
13. bacteria
14. eukaryotes
15. Eukarya
16. protists
17. protozoans
18. algae
19. photosynthesis
20. fungi
21. by breaking them down with digestive juices
22. Answers may vary. Sample answer: molds and mushrooms
23. Members of kingdom Plantae are green and multicellular, have cell walls, and make their own food using the sun's energy.
24. sunlight
25. land, water
26. Answers may vary. Sample answer: The sugar, or glucose, that plants make for themselves is also important to other organisms because some, such as animals, eat the plants and others, such as fungi, absorb and digest dead plant material that contains the glucose.
27. Besides making food for other organisms, plants such as trees, grasses, and bushes can be habitats for birds, insects, and other animals.
28. They are multicellular and can move.
29. environment
30. animals
31. Animals need plants for food, and some use them for habitat.
32. Animals depend on bacteria and fungi to recycle nutrients found in dead organisms.
33. sponges

Vocabulary and Section Summary A

SECTION: SORTING IT ALL OUT

1. classification: the division of organisms into groups, or classes, based on specific characteristics
2. taxonomy: the science of describing, naming, and classifying organisms

SECTION: DOMAINS AND KINGDOMS

1. Archaea: in a modern taxonomic system, a domain made up of prokaryotes that differ from other prokaryotes in the makeup of their cell walls and in their genetics
2. Bacteria: in a modern taxonomic system, a domain made up of prokaryotes that differ from other prokaryotes in the makeup of their cell walls and in their genetics
3. Eukarya: in a modern taxonomic system, a domain made up of all eukaryotes
4. Protista: a kingdom of mostly one-celled eukaryotic organisms that are different from plants, animals, bacteria, archaea, and fungi
5. Fungi: a kingdom made up of nongreen, eukaryotic organisms that have no means of movement, reproduce by using spores, and get food by breaking down substances in their surroundings and absorbing the nutrients
6. Plantae: a kingdom made up of complex, multicellular organisms that are usually green, have cell walls made of cellulose, cannot move around, and use the sun's energy to make sugar by photosynthesis
7. Animalia: a kingdom made up of complex, multicellular organisms that lack cell walls, can usually move around, and quickly respond to their environment

Vocabulary and Section Summary B

SECTION: SORTING IT ALL OUT

1. taxonomy
2. systematics
3. classification
4. branching diagram
5. shared derived characteristics

Levels of Classification

a. domain
b. kingdom
c. phylum
d. class
e. order
f. family
g. genus
h. species

SECTION: DOMAINS AND KINGDOMS

1. Bacteria
2. Plantae
3. Archaea
4. Animalia
5. Fungi
6. Protista
7. Eukarya

D	I	C	H	O	T	A	Y	R	A	K	U	E	Y	I
Y	A	L	S	Y	A	R	Z	A	U	R	Q	W	F	J
G	F	A	S	P	X	E	S	K	P	P	X	B	I	U
P	K	S	O	H	O	X	A	N	S	S	K	A	F	Y
E	E	S	F	U	N	G	I	I	N	W	I	C	A	Y
C	G	I	Y	G	O	F	T	P	L	A	N	T	A	E
P	D	F	H	X	M	K	F	L	K	A	S	E	C	Q
W	M	I	G	H	Y	N	F	O	Z	I	M	R	F	P
A	R	C	H	A	E	A	A	C	T	E	R	I	A	M
P	G	A	R	L	R	W	S	O	F	S	R	A	N	I
V	U	T	J	X	I	F	R	L	B	V	N	Y	F	A
J	W	I	S	T	W	P	U	O	A	F	N	L	F	D
D	S	O	Y	G	E	O	I	X	Y	R	A	C	K	K
S	F	N	E	K	D	W	D	X	Q	H	Z	T	S	C

Reinforcement

KEYS TO THE KINGDOM

Animalia:

Complex organisms with no cell walls
Have specialized sense organs
Depend on plants for food

Plantae:

Usually green
Make food through photosynthesis
Use the sun's energy to make sugar
Giant sequoias

Protista:

Most are single-celled or simple multicellular organisms.
Algae
Protozoans
Evolved from bacteria about 2 billion years ago

Fungi:

Break down material outside their bodies and then absorb the nutrients
Do not perform photosynthesis
Molds
Mushrooms

Critical Thinking

1. The Noid and the Druff clans cannot communicate knowledge about plant and animal life because they have different names for plants and animals. This is a problem if they want to work closely together as allies.
2. The clans have developed different knowledge of plants and animals over time, and they use different common names that reflect that knowledge.
3. The clans cannot easily inform each other about the dangers or usefulness of plants or animals. As a result, clan members are not trading information that could prevent problems or present solutions in the future.
4. Answers may vary. Sample answer: No, I don't think this would be a good idea. Although the clans would probably be able to learn the names quicker this way, the Noid names would not have any significant meaning for the Druffs, and vice versa. The names would therefore be easy to forget and would not communicate any useful information.
5. Answers may vary. Sample answer: Perry Noid should urge the clans to keep their common names for plants and animals. But, he should also encourage them to create a scientific name for each species. The scientific name should have a clear meaning that both clans can understand.

SciLinks Activity

Answers may vary. Sample illustrations:

1. a mushroom growing on a forest floor, where there is little sunlight, surrounded by rotting leaves, decomposing sticks, and a crumbling log
2. a sunflower growing along a fence line with a stream nearby where there is a lot of bright sunlight
3. a horse eating grass in a pasture, or a bird building a nest out of twigs on the branch of a tree

Section Review

SECTION: SORTING IT ALL OUT

1. Sample answer: Classification is putting things into orderly groups based on similar characteristics. Taxonomy is the science of describing, classifying, and naming organisms.
2. Sample answer: Scientists use scientific names for organisms to make it easier for people all over the world to understand which organism is being referred to.
3. Domain, kingdom, phylum, class, order, family, genus, and species.
4. Moss is the oldest kind of organism. The hibiscus or the pine tree may be considered the youngest kinds of organisms.
5. The fern, pine tree, and hibiscus have tissues that transport materials.
6. Sample answer: I would make a list of the new organism's characteristics.
7. Sample answer: Organisms that have many shared derived characteristics are more closely related to each other than organisms that do not have many shared derived characteristics are.
8. Sample answer: I would infer that the species at the tip of the branch evolved from the species that occurs halfway up the branch.
9. Sample answer: I would infer that species at the tips of branches that did not reach the top of the branching diagram were extinct.

SECTION: DOMAINS AND KINGDOMS

1. Sample answer: Archaea is a domain made up of prokaryotes that are often found in extreme environments. Bacteria is a domain made up of prokaryotes that usually live in more moderate environments.
2. Sample answer: All members of the domain Eukarya have a nucleus and membrane-bound organelles in their cells. The kingdom Protista includes eukaryotes that don't fit into the other kingdoms. The kingdom Fungi includes eukaryotic organisms that make spores, don't make their own food, and don't move from place to place. The kingdom Plantae includes eukaryotes that perform photosynthesis and have cells with cell walls. The kingdom Animalia includes multicellular eukaryotes that have cells without cell walls, can usually move from place to place, and have specialized sense organs.
3. Sample answer: All organisms in the three domains are made up of cells.
4. Sample answer: The group of organisms evolved after birds, but before mammals.
5. Sample answer: I would classify this organism in the domain Archaea, because the domain Archaea includes prokaryotes that live in extreme environments.
6. After 5 hours; The starting population doubles every 30 min, therefore $1 \times 2 = 2$; $2 \times 2 = 4$; $4 \times 2 = 8$; $8 \times 2 = 16$; $16 \times 2 = 32$; $32 \times 2 = 64$; $64 \times 2 = 128$; $128 \times 2 = 256$; $256 \times 2 = 512$; $512 \times 2 = 1024$
7. Sample answer: Scientists use a classification system to group these organisms to help us understand the relationships between organisms.
8. Sample answer: The Venus' flytrap should be in the kingdom Plantae because it has cell walls. Its ability to consume insects makes it an unusual member of the kingdom Plantae.

Chapter Review

1. B
2. Archaea
3. Animalia
4. prokaryotes
5. A
6. A
7. Sample answer: Each species is unique, and scientific names make it possible for scientists all over the world to know specifically which organism is being discussed without the confusion of common names.
8. Taxonomists classify organisms based on the shared derived characteristics of the organisms.
9. no; A bacterium is a prokaryote because it does not have a nucleus.
10. lemur
11. chimpanzee
12. no; Lemurs branched off between points A and B.
13. Answers may vary; accept all reasonable responses. Students should be able to articulate that not all organisms are either plants or animals. Some students may even point out that some organisms have characteristics of both plants and animals, and therefore could not be classified as either a plant or an animal.
14. An answer to this exercise can be found a the end of the Teacher Edition.
15. Sample answer: Each level of classification groups organisms according to the characteristics that they share. At broader levels of classification, such as domain and kingdom, members of the group share fewer characteristics than they do at more specific levels, such as family and genus.
16. Sample answer: All members of the four kingdoms in the domain Eukarya are made up of eukaryotic cells. The DNA of these organisms are found in the nucleus in their cells.
17. Sample answer: I would study the characteristics of the fossil and compare them to what I know about fish, amphibians, reptiles, birds, and mammals. Then, I would determine which group of organisms the fossil is most closely related to and determine the fossil's place on the branching diagram.
18. Sample answer: The organism is probably extinct and no other organisms have been found that seem to have developed from the extinct organism.
19. Sample answer: One way to determine into which domain a prokaryote should be classified is to know what type of environment the prokaryote lives in.
20. Sample answer: no; Sam should first find out if there are other organisms with similar characteristics to determine whether the new organism may be related to other organisms.
21. Sample answer: The organism has long, leafy appendages and a tail.
22. Sample answer: I think this organism is an animal because it has a mouth and a tail. However, I can't see enough characteristics to decide what kind of animal it is.
23. Sample answer: Other useful information might include where the organism lives, how the organism breathes, and how the organism moves.
24. The difference is 278° C;
    $536°F = 5/9 \times (536 - 32) = 280°C$;
    $35.6°F = 5/9 \times (35.6 - 32) = 2°C$;
    $280°C - 2°C = 278°C$
25. Sample answer: Members in each of the other kingdoms share characteristics that allow them to be grouped together. Kingdom Protista is unusual because it's members are grouped together not because they have characteristics in common but because they d not fit in the other kingdoms.

Section Quizzes

SECTION: SORTING IT ALL OUT

1. B
2. A
3. B
4. C
5. D
6. A

SECTION: DOMAINS AND KINGDOMS

1. B
2. A
3. C
4. F
5. D
6. E
7. A
8. B
9. C

Chapter Test A

1. C
2. B
3. C
4. D
5. C
6. B
7. A
8. B
9. A
10. D
11. C
12. B
13. A
14. B
15. C
16. A
17. D
18. genera
19. classes
20. domains
21. orders
22. families
23. species

Chapter Test B

1. B
2. A
3. A
4. C
5. D
6. C
7. A
8. A
9. C
10. D
11. B
12. C
13. D
14. B
15. A
16. F
17. C
18. D
19. B
20. E
21. A
22. G
23. C
24. A
25. C

Chapter Test C

1. taxonomy
2. branching diagram
3. Protista
4. Plantae
5. Animalia
6. Archaea
7. D
8. C
9. B
10. A
11. C
12. Before the time of Carolus Linnaeus, some organisms had names that were up to 12 words long. In addition, many organisms had more than one common name. Both these situations made communicating about certain living things complicated. Creating a scientific name for organisms by using the genus and specific name simplified communication and also gave each organism just one name.
13. Scientists divide prokaryotes into groups based on the differences of their genetics, the makeup of their cell wall, and their ability (or inability) to live in extreme environments.

14. A sponge is classified as a simple animal and not a plant, because even though it does not move, it must get its food from its environment and cannot make food through photosynthesis. In addition, sponges do not have cell walls or chlorophyll.
15. Answers may vary. Sample answer: When scientists did not have microscopes, they could not learn about microscopic organisms. These organisms would have included many that caused disease and decomposition in humans and other animals. Not knowing about certain disease-causing bacteria would make it difficult for doctors to treat the illnesses properly, and so more people, farm animals, and pets would get ill and possibly die of these diseases than do now.
16. Answers may vary. Sample answer: Earth may be colder than it is today, with many organisms having become extinct due to the cold. However, many archaea will be the same because they can handle extreme environments. Humans might develop more hair to stay warmer. A new organism on Earth might be a fungus-like organism that is able to move itself from place to place across the cold landscape in order to find materials it can absorb and digest. Unlike animals, fungi would have limited ways of responding to changes in their environment.
17. Species *A* through *I* would all share the same first word in their scientific names. For example, if they were all cats, their scientific names would all begin with Felis followed by a unique specific name.
18. *I* and *J* are on different sides of the first big branch. This means they are in distantly related groups. *I* and *A* are in the same group, so they are more closely related than *I* and *J*.
19. **a.** domains, **b.** Eukarya, **c.** kingdoms, **d.** Protista

Performance-Based Assessment

1. Answers may vary. Sample answer: A collection of buttons can be put into "classes" if we first look at the buttons carefully and then decide which characteristics some buttons share and which characteristics make other buttons distinct.
1. Answers may vary. Sample answer: I have three classes of buttons.
2. Answers may vary. Sample answer: Class 1: Duodicus – buttons with two holes; Class 2: Quadrophenius – buttons with four holes; Class 3: Solidicus – buttons with no holes
3. Answers may vary. Sample answer: No, all my buttons fit into at least one of the classes.
4. Answers may vary. Sample answer: I have four families of buttons.
5. Answers may vary. Sample answer: Rojo – red buttons; Blanco – white buttons; Azul – blue buttons; Anaranjado – orange buttons
6. The next classification level is called *genus*.
7. Answers may vary. Sample answer: The next division could be decided by button size or thickness.
8. Answers may vary. Sample answer: Like organisms, the buttons have been classified by like characteristics. As greater numbers of different kinds of buttons were classified, my classification system was adjusted to accomodate new button characteristics, just as scientists adjust the classification system as new types of organisms are discovered.

Explore Activity

DATASHEET A

3. the age of the person wearing those clothes.
4. teenager
5. infant, toddler, and preteen
6. Sample Answer: Teenager clothing may be divided into radical and conservative kinds of clothing.

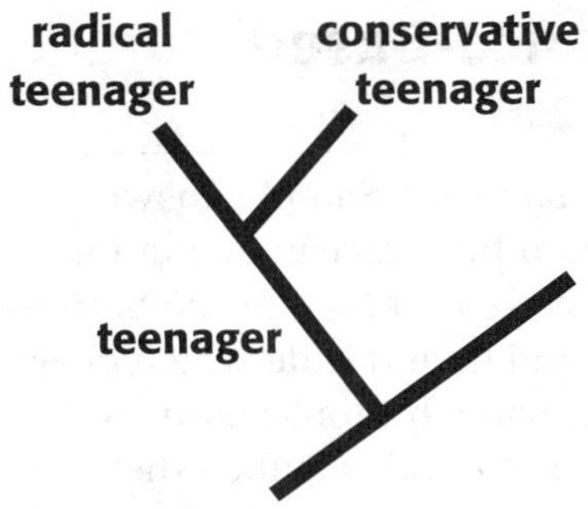

DATASHEET B

3. Sample answer: This diagram shows how clothing can be divided into groups based on the age of the person wearing the clothes.
4. teenager
5. infant, toddler, and preteen
6. Sample answer: Teenager clothing may be divided into radical and conservative kinds of clothing.

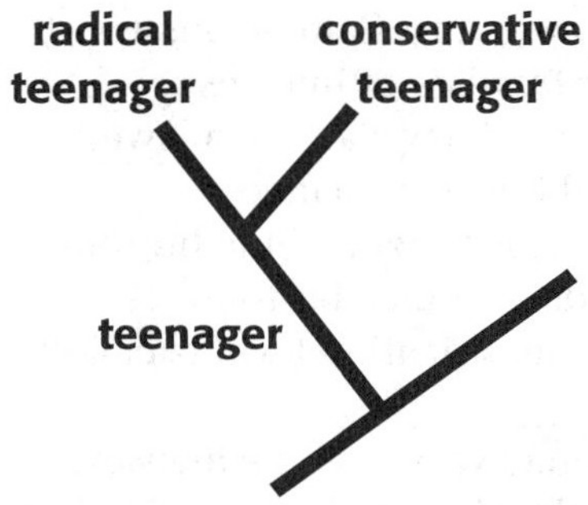

DATASHEET C

3. Sample answer: This diagram shows how clothes can be divided into groups based on the age of the person wearing those clothes.
4. teenager
5. infant, toddler, and preteen
6. Sample Answer: Teenager clothing may be divided into radical and conservative kinds of clothing.

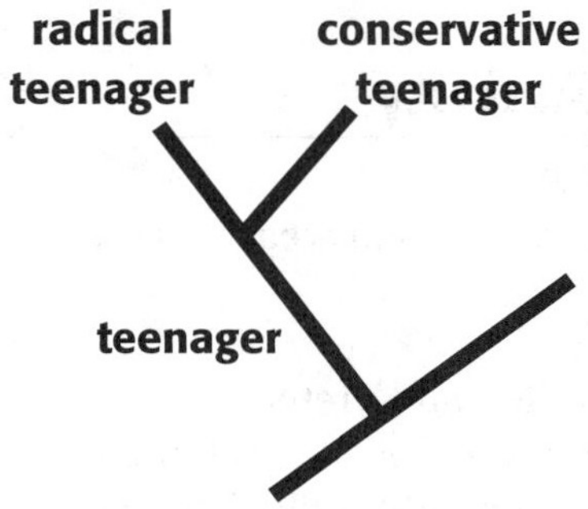

Quick Lab: Constructing a Branching Diagram

DATASHEET A

1.–3. Sample answer shown below.

4. Rabbits are more closely related to kangaroos than to frogs and snakes because rabbits and kangaroos share more derived characteristics than rabbits and frogs or rabbits and snakes do.

DATASHEET B

1.-3. Sample answer shown below.

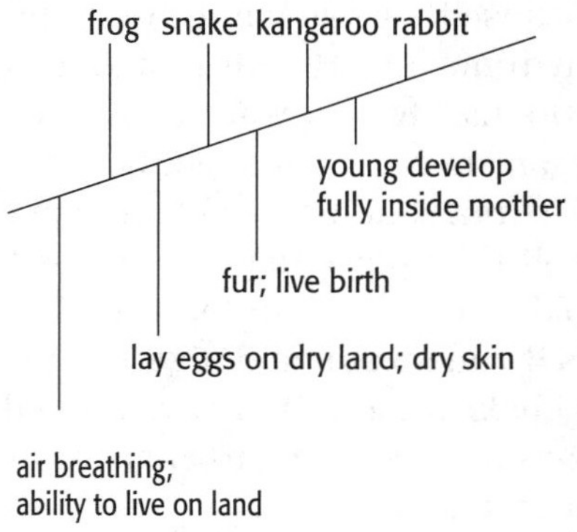

4. The diagram shows how the organisms are related by the kinds of characteristics that they share.

DATASHEET C

1.–3. Sample answer shown below.

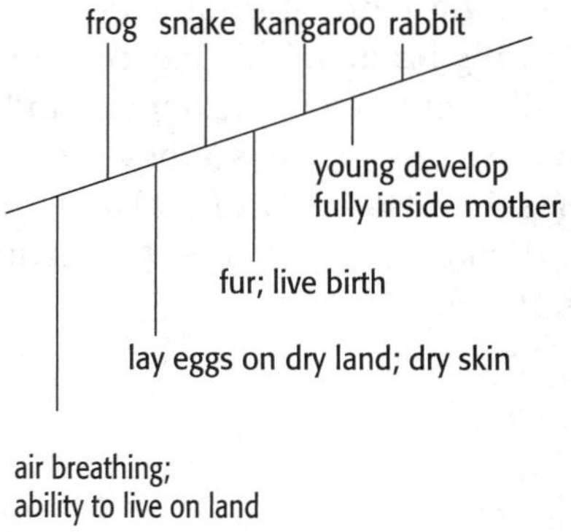

4. The diagram shows how the organisms are related by the kinds of characteristics that they share.

Quick Lab: Fossils and Branching Diagrams

DATASHEET A

2. yes; yes; answers may vary.

3.–4.

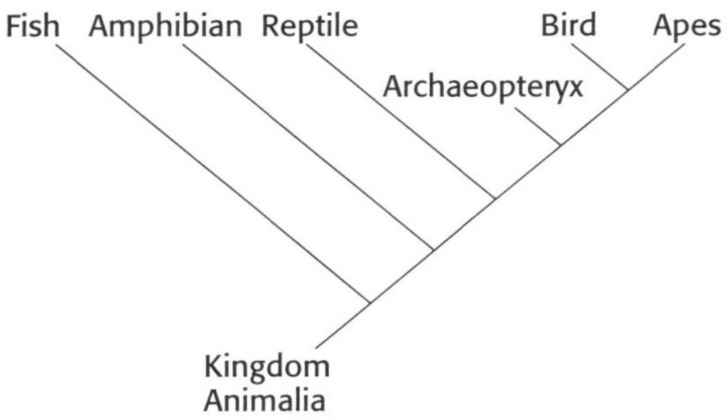

5. The branch with archaeopteryx at its tip does not reach the top of the diagram.
6. Birds and apes evolved more recently than the archaeopteryx evolved. Fish, amphibians, and reptiles evolved before the archaeopteryx.

DATASHEET B

2. Sample answer: The archaeopteryx has a backbone, legs, and wings with feathers.

3.–4.

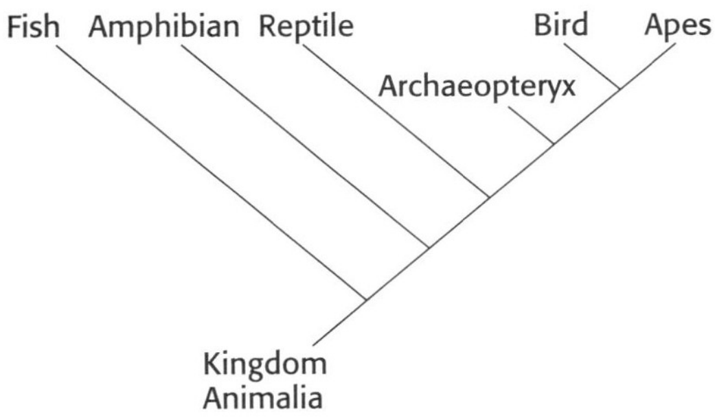

5. The branch with archaeopteryx at its tip does not reach the top of the diagram.
6. Birds and apes evolved more recently than the archaeopteryx evolved. Fish, amphibians, and reptiles evolved before the archaeopteryx.
Sample answer: I know that the archaeopteryx evolved after fish, amphibians, and reptiles because it has a characteristic they lack: wings with feathers.

DATASHEET C

2. Sample answer: The archaeopteryx has a backbone, legs and wings with feathers.

3.–4.

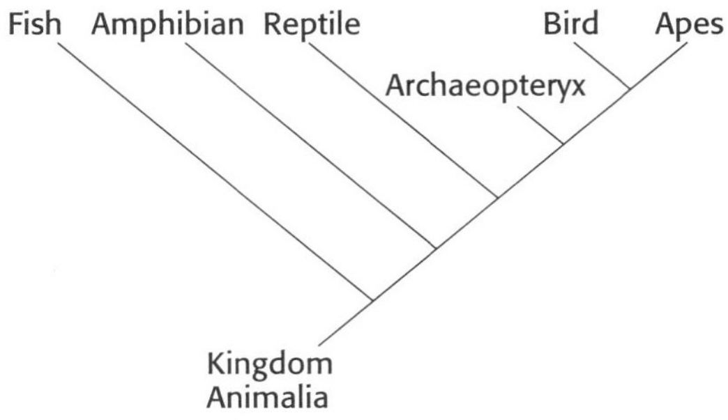

5. The branch with archaeopteryx at its tip does not reach the top of the diagram.
6. Birds and apes evolved more recently than the archaeopteryx evolved. Fish, amphibians, and reptiles evolved before the archaeopteryx.
Sample answer: I know that the archaeopteryx evolved after fish, amphibians, and reptiles because it has a characteristic they lack: wings with feathers.

Chapter Lab

DATASHEET A

4.

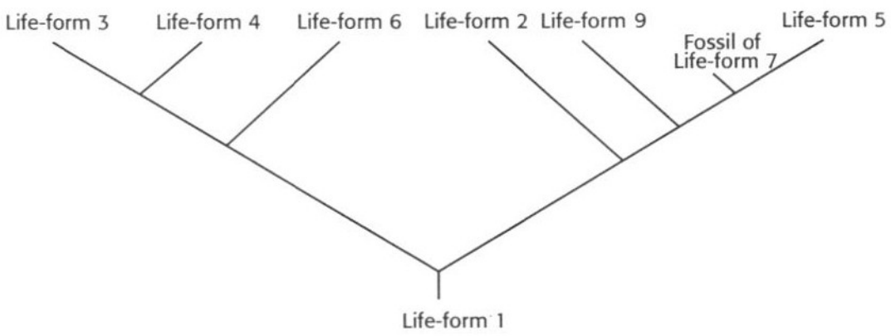

7. Sample answers: Life-forms in room 1 have more than one repeating body segment.
The shapes of these repeating segments differ between life-forms.
8. Sample answers: The bodies of life-forms in room 2 are not divided into repeating body segments.
Life-forms in room 2 differ from the life-forms in room 1 because of the number of repeating body segments.
9. The fossil of life-form 7 shows that the body of life-form 7 is not divided into repeating body segments. Therefore, life-form 7 is likely to be most closely related to life-form 2 and life-form 5.
10. Sample answer: I placed life-form CC9 in room 2 because the body of life-form CC9 is not divided into repeating body segments.
11. Sample answers: Yes, these other characteristics could be used to make a branching diagram.
Using different characteristics would change the branching diagram.
It would be hard for scientists to decide which characteristics to use and the scientists might not agree with each other.
12. Living organisms and extinct organisms can be grouped based on their shared derived characteristics.
Branching diagrams show the relationships between living organisms and extinct organisms along a timeline.

DATASHEET B

4.

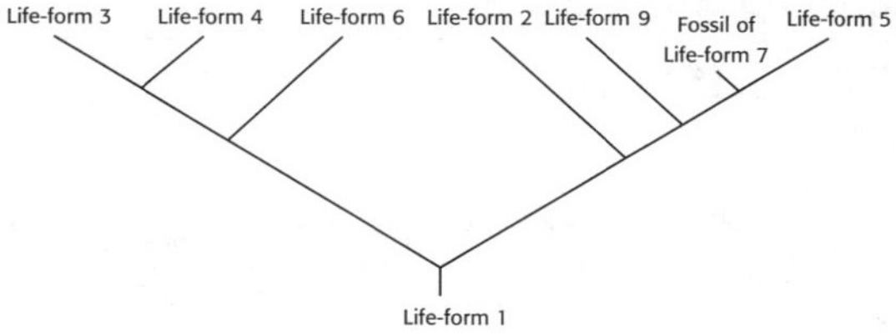

7. Sample answer: Life-forms in room 1 have more than one repeating body segment. The shapes of these repeating body segments differ between life-forms.

8. Sample answer: The body of each life-form in room 2 is shaped differently. The bodies of life-forms in room 2 are similar to each other because they are not divided into repeating body segments. The lifeforms in room 2 differ from the life-forms in room 1 because of the number of repeating body segments.
9. The fossil of life-form 7 shows that the body of life-form 7 is not divided into repeating body segments. Therefore, life-form 7 is likely to be most closely related to life-form 2 and life-form 5.
10. Sample answer: I placed life-form CC9 in room 2 because the body of life-form CC9 is not divided into repeating body segments.
11. Sample answer: Other characteristics that might have been used to classify these different life-forms include the number and kind of eyes the life-forms have, whether or not the life-forms have legs, and whether or not the life-forms have mouths. Using different characteristics would change the branching diagram. It would be hard for scientists to decide which characteristics to use and the scientists might not agree with each other.
12. Living organisms and extinct organisms can be grouped based on their shared derived characteristics. Branching diagrams show the relationships between living organisms and extinct organisms along a timeline.

DATASHEET C

4.

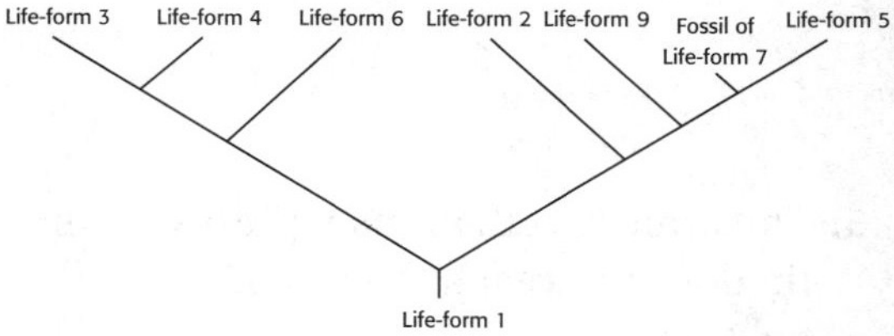

7. Sample answer: Life-forms in room 1 have more than one repeating body segment. The shapes of these repeating body segments differ between life-forms.

8. Sample answer: The body of each life-form in room 2 is shaped differently. The bodies of life-forms in room 2 are similar to each other because they are not divided into repeating body segments. The life-forms in room 2 differ from the life-forms in room 1 because of the number of repeating body segments.
9. The fossil of life-form 7 shows that the body of life-form 7 is not divided into repeating body segments. Therefore, life-form 7 is likely to be most closely related to life-form 2 and life-form 5.
10. Sample answer: I placed life-form CC9 in room 2 because the body of CC9 is not divided into repeating body segments.
11. Sample answer: Other characteristics that might have been used to classify these different life-forms include the number and kind of eyes the life-forms have, whether or not the life-forms have legs, and whether or not the life-forms have mouths. Using different characteristics would change the branching diagram. It would be hard for scientists to decide which characteristics to use and they might not agree with each other.
12. Living organisms and extinct organisms can be grouped based on their shared derived characteristics. Branching diagrams show the relationships between living organisms and extinct organisms along a timeline.

Science Skills Activity

DATASHEET

4. Accept all reasonable answers. Students should be able to articulate whether their report is objective.
5. Sample answer: I assumed that all of the information that I collected was accurate. My conclusions may be wrong if the information that I used is inaccurate.